Optical Fiber
Transmission Systems

Applications of Communications Theory
Series Editor: R. W. Lucky, *Bell Laboratories*

A Continuation Order Plan is available for this series. A continuation order will bring delivery of each new volume immediately upon publication. Volumes are billed only upon actual shipment. For further information please contact the publisher.

Optical Fiber Transmission Systems

Stewart D. Personick

TRW Vidar Division
Mountain View, California

PLENUM PRESS · NEW YORK AND LONDON

813547

Library of Congress Cataloging in Publication Data

Personick, Stewart D
  Optical fiber transmission systems.

  (Applications of communications theory)
  Includes index.
  1. Optical communications. 2. Fiber optics. I. Title. II. Series.
TK5103.59.P47   621.36'92   80-20684
ISBN 0-306-40580-6

First Printing — February 1981
Second Printing — January 1983

© 1981 Plenum Press, New York
A Division of Plenum Publishing Corporation
233 Spring Street, New York, N.Y. 10013

Printed in the United States of America

I will always be indebted to the men and women who have taught me, encouraged me, guided me — and inspired me — throughout my career. This book is dedicated to them.

Preface

In the 14 years which have passed since the first proposal that ultra-low-loss optical fibers could in principle be fabricated, astounding progress has taken place in this new technology.[1] What was only an interesting possibility in the late sixties was already a practical reality for several applications in the late seventies.[2-5] The score card for innovation is impressive. Attenuations in fibers have been reduced from decibels per meter to decibels per kilometer in practical cables. Projected lifetimes of lasers have been increased from minutes to decades at room temperature. Practical techniques for interconnecting fibers with micrometer tolerances have been demonstrated in the field. Fibers which at one time broke spontaneously can be manufactured in kilometer lengths with strengths exceeding that of fine steel.

Fiber optics for transmission is a broad subject. It can be viewed from many different perspectives: e.g., materials, devices, theory, subsystems, systems, applications, etc.

Several books and hundreds of articles have been written on this subject. No doubt more will follow. Many of these touch on the overall subject, but like this book they emphasize one of these perspectives. This book is a systems engineer's view of fiber optics for transmission. It presents the author's viewpoint, which evolved over a period of 10 years while he participated in research, exploratory development, development, and finally the applications of fiber transmission systems.

As much as possible, the author has avoided reproducing long derivations of analytical results, so long as they were available in easy-to-obtain references. Results, heuristic arguments to motivate those results, and references for the details of the derivations have been used frequently throughout.

In the applications section (Chapter 5) the author has tried to explain in each instance what the application is, what the requirements are, and what competitive technologies are available. Detailed cost comparisons were avoided because of the rapid changes occurring in fiber technology.

I would like to express my sincere appreciation to Genevieve Van Dera and Edith Coon, who prepared the manuscript; and to Christopher Kent and John Schultz, who prepared the illustrations and photographs.

Mountain View, California   Stewart D. Personick

Contents

Fundamentals

1.1. Block Diagram of a Typical Fiber System

A typical digital fiber optic transmission application is illustrated in the block diagram of Figure 1.1. Starting at the upper left, we assume that the information to be transmitted originates in analog form, as, for example, in 4-kHz telephone channels. Several such analog signals can be converted simultaneously to individual sequences of binary digits, each at a fixed data rate (bits/s), using a piece of standard electronic terminal equipment known as a channel bank. Such conversion from analog to digital form is done routinely in transmission systems using twisted pairs, coaxial cables, radio, etc. These individually encoded voice signals are then inter-leaved to form a single (combined) time-multiplexed digital signal which can be used to modulate an optical source.[1] There are two ways to perform the modulation.[2] One approach is to generate a continuous optical signal and then modulate that signal in amplitude or phase using an optical modulator following the source. Another approach is to modulate the electrical power driving the source in order to modulate the source output power directly. The former approach is generally more expensive and difficult. External modulation typically would be used only when it is impractical to directly modulate the source. One limit on direct modulation is speed. Some sources respond very slowly to the variations in their electrical drive power. Fortunately, the main candidates for optical sources in fiber transmission applications—injection lasers and light-emitting diodes— can be modulated directly at rates suitable for many applications, making simple inexpensive optical transmitters possible. A more complete discussion of modulation will be given in Sections 2.2 and 3.3. Present widely available sources for fiber applications are made of gallium aluminum arsenide and emit in the 0.75–0.9-μm region of wavelengths. Sources in the 1–1.5-

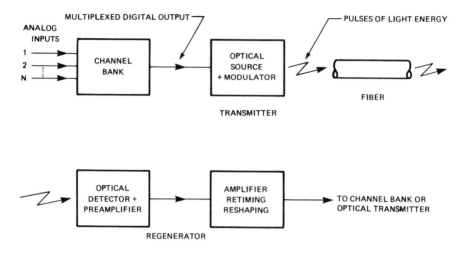

Figure 1.1. Typical digital system.

μm region made from GaInAsP are being developed, but are not yet widely available.

The modulated light power emitted by the optical transmitter is coupled to an optical fiber for transmission.[3] Typically the fiber is simply brought into close proximity to the light-emitting surface of the source. One tries to make the source emission area comparable to the fiber cross section in order to avoid costly and impractical coupling optics. Propagation in fibers will be described in more detail in Sections 1.2, 3.5, and 4.1. The basic propagation mechanism is total internal reflection.[4] The fiber consists of a core region and a cladding region as shown in Figure 1.2. The index of refraction of the cladding material (typically glass or plastic) is slightly lower than the index of refraction of the core material. Typical fiber dimensions are a 50–100-μm diameter for the core and a 100–200-μm diameter for the cladding (although other sizes are sometimes appropriate, as will be discussed in Sections 4.1.4 and 4.2.2). Fibers with losses of less than 10 dB/km allow transmission of the light pulses over several kilometers distance before a "regenerator" is required. Cables with less than 4 dB/km loss at 0.8-μm wavelength are commercially available. The regenerator consists of a detector (typically a photodiode made from silicon) which converts the light pulses (now weakened and possibly distorted by propagation) into electrical pulses, electronic amplifiers to increase the level of the detector output pulses, and electronic retiming and reshaping circuits to generate new, undistorted electrical pulses. These new pulses can drive

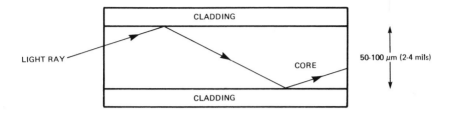

Figure 1.2. Total internal reflection.

a terminal to recover the original analog messages or can drive another optical transmitter for further transmission.

The system as described above is fairly simple. In many ways it is more like a wire (twisted pair or coaxial) communication system (using photons instead of electrons) than it is like an optical frequency extension of radio. There were, however, a number of serious hurdles which had to be overcome before fiber systems could become a commercial reality. The following subjects are treated in the sections below: cabling and splicing fibers; producing inexpensive low-loss, high-bandwidth, high-strength fibers; producing low-cost high-reliability sources and detectors, at the appropriate wavelengths, with the appropriate input–output characteristics; and designing receiver and transmitter subsystems for easy interfacing with conventional electronic equipment.

1.2. Fibers

1.2.1. Glass and Fiber Making†

Most low-loss optical fibers are made of what is usually called glass. To most people, glass is an amorphous solid with a high percentage of silicon dioxide. Actually a glass can be composed of any of a number of "glass-forming" materials which under proper conditions of preparation will form an amorphous rather than a crystalline solid. For example, solid water is usually crystalline (ice), but under unusual fast-cooling conditions it can be a glass. Most glasses used in the early days of fiber optics research were mixtures of sodium oxide, calcium oxide, and silicon dioxide— known as soda–lime–silicate glasses (in recognition of the starting materials from which they are made). Such glasses have a relatively low melting point, and indeed can be melted in platinum or fused-quartz crucibles.

† See Chapters 7–9 of Reference 4.

This property led to their choice in early research. It is difficult to fabricate such multicomponent glasses without including impurities in the form of various metallic ions (Fe, Cu, Ni, etc). These impurities must be reduced to concentrations of a few parts per billion, if the attenuation of the glass is to be less than 10 dB/km at 0.9-μm wavelength. In early fibers having losses of 100 dB/km or more, absorption by such metallic ion impurities was the limiting factor. [Light which is absorbed is converted to heat (vibrational energy) and reradiated at long wavelengths, or conducted away.] A major breakthrough in fiber technology occurred in 1970 when Corning Glass Works fabricated fibers of essentially pure fused silica (SiO_2) with losses below 20 dB/km. The starting materials (silicon and germanium tetrachloride) and the fabrication process (chemical vapor deposition, CVD) allowed for the the fabrication of fibers with the required low concentrations of impurities. The method used by Corning is shown in Figure 1.3. $SiCl_4$ and $GeCl_4$ were burned in oxygen to form fine particles of SiO_2 and GeO_2 which were deposited as "soot" on a rotating carbon mandrel. By increasing the amount of GeO_2 one could raise the refractive index of the deposited material. Thus GeO_2 was included in the initially deposited layers, which eventually formed the higher-index fiber core. When a sufficiently large amount of material was deposited, the mandrel was removed and the tube of glass soot was "sintered" (consolidated) into a clear glass "preform" by heating. Further heating caused the preform to collapse into a solid rod. This rod could be made into a fiber by using the equipment shown in Figure 1.4.

The preform is lowered into a donut-shaped oven which raises its temperature to the softening point (since glass does not make an abrupt transition from solid to liquid, this temperature is not exactly defined). The softened part of the preform is then pulled down (like taffy) from the end and attached to a rotating drum. The resulting fiber can be "pulled"

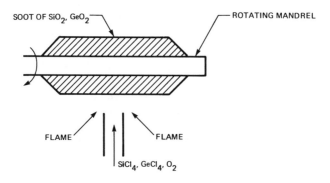

Figure 1.3. CVD process.

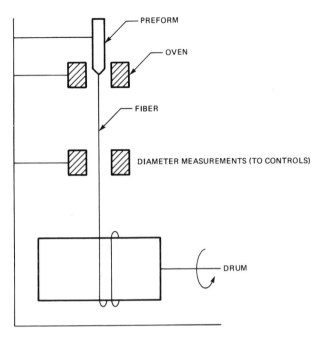

Figure 1.4. Fiber drawing machine.

with a diameter which depends upon the oven temperature, the rotational speed of the drum, and the rate at which the preform is lowered into the oven. Fiber-diameter-measuring instruments allow these parameters to be feed-back controlled.

An alternative way to fabricate high-silica preforms (the modified chemical vapor deposition, MCVD, process) is shown in Figure 1.5. One begins with the tube of pure fused silica, which is inserted into a glass-working lathe. The lathe allows gases to flow into one end of the tube and out the other, while the tube rotates between synchronous chucks. Mixtures of germanium tetrachloride, silicon tetrachloride, and boron trichloride (as well as other possible materials) are burned in oxygen inside the tube owing to the presence of a traveling torch outside of the tube. The resulting oxides deposit downstream from the torch in the form of soot. The soot is consolidated into a clear glass as the torch passes over it. In this way layer upon layer of new glass can be deposited inside the tube, with varying compositions and thus a varying index of refraction. After hundreds of layers are deposited, the temperature of the torch is raised, and the tube collapses to a solid preform rod.

Various other methods of preform fabrication and fiber drawing can be found in the literature for both high-silica and multicomponent glasses.

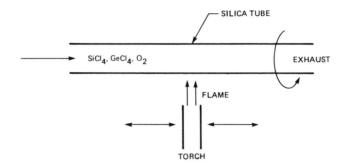

Figure 1.5. MCVD process.

Figure 1.6 shows the loss vs. wavelength characteristic of a typical commercially available low-loss fiber. Figure 1.7 shows how the best reported fiber loss at 820-nm wavelength has improved with time since the late sixties. In early fibers the loss was dominated by absorption due to the presence of impurities. In the best modern fibers the loss at 0.8–0.9-μm wavelength and at 1–1.5-μm wavelength is dominated by scattering caused by molecular density and compositional fluctuations. This is known as Rayleigh scattering. There is usually a residual "water" absorption peak near 0.9-μm wavelength owing to the presence of residual OH ions in the glass. Since Rayleigh scattering decreases as the fourth power of the wavelength, losses are lower in the 1.3–1.5-μm region.

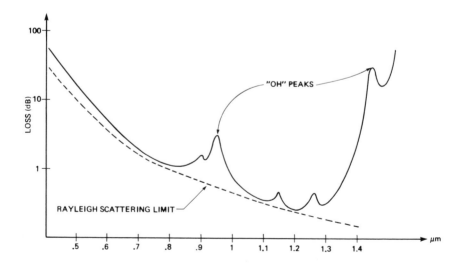

Figure 1.6. Loss vs. wavelength. Typical low-loss silica fiber.

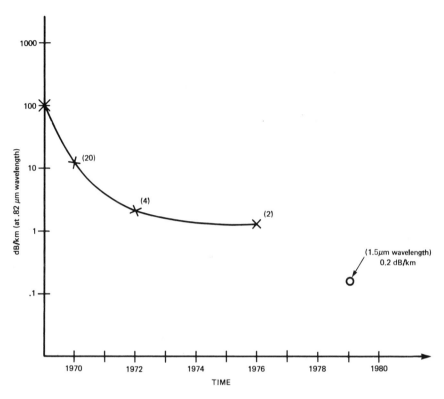

Figure 1.7. Best reported loss vs. time.

In addition to drawing fibers with well-controlled parameters (e.g., diameter, refractive index), one must be concerned with coating fibers to preserve their strength and with forming coated fibers into practical cables. These subjects are beyond the scope of this book, but are treated thoroughly in the references.†

1.2.2. Propagation‡

A number of different types of fibers are currently of interest for various applications. The simplest fiber type consists of a core of one type of glass inside a cladding of glass or plastic having a lower index of refraction than the core. Such a fiber is illustrated in Figure 1.8. Typically the diameter of the core is tens or hundreds of times the wavelength of the propagating

---

† See, e.g., Chapters 10 and 13 of Reference 4.
‡ See Chapters 3, 4, 6, and 11 of Reference 4.

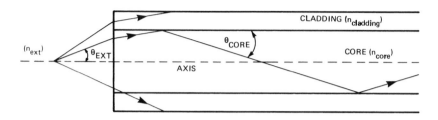

Figure 1.8. Total internal reflection, critical angle.

light. Because of this, the propagation phenomena can be studied using geometrical optics rather than wave optics, if desired. A typical core diameter might be 50 μm or 2 mils. As mentioned above typical wavelengths are 0.75–1.5 μm. Propagation in the fiber can be studied using the ray model shown in Figure 1.8. Rays from a light source which strike the fiber are refracted (bent) because of the index of refraction difference of air and glass. Typical glasses have an index of refraction of about 1.5. Rays which enter the core and which strike the core–cladding interface at a sufficiently shallow angle are totally internally reflected. The maximum angle which a totally reflected ray may have relative to the fiber axis, the "critical angle," is obtained using Snell's law of refraction of a *transmitted* ray striking the core–cladding interface and taking the limiting case of having the transmitted ray emerge parallel to the interface:

$$n_{core} \cos(\theta_{core}) = n_{cladding} \cos(\theta_{cladding}) \mid \cos(\theta_{cladding}) = 1 \qquad (1.2.1)$$

(no transmission through the interface), i.e.,

$$n_{core} \cos(\theta_{max}) = n_{cladding}$$

Therefore

$$\sin \theta_{max} = \left[1 - (n_{cladding}/n_{core})^2\right]^{1/2}$$

where n_{core} is the index of refraction of the core, $n_{cladding}$ is the index of refraction of the cladding, and θ_{max} is the critical angle.

If we assume that the index of refraction difference between core and cladding is small, as is often the case, then we obtain the following relationship:

$$n_{core} \sin(\theta_{max}) = (n_{core}^2 - n_{cladding}^2)^{1/2}$$
$$= \left[(n_{core} + n_{cladding})(n_{core} - n_{cladding})\right]^{1/2}$$

If

$$n_{core} \simeq n_{cladding} = n$$

then

$$n \sin(\theta_{\max}) \cong n(2\Delta)^{1/2} = \text{N.A.} \qquad (1.2.2)$$

where $\Delta = (n_{\text{core}} - n_{\text{cladding}})/n_{\text{core}}$ is the "index step" and N.A. is the numerical aperture of the fiber.

Equation (1.2.2) gives the maximum angle which a ray may have relative to the fiber axis inside the fiber. One can also derive the maximum angle which a ray incident upon the fiber end can have relative to the fiber axis in the medium external to the fiber. Again from Snell's law we obtain:

$$n_{\text{ext}} \sin(\theta_{\text{ext}}) = n \sin(\theta_{\text{core}}) \qquad (1.2.3)$$

$$n_{\text{ext}} \sin(\theta_{\text{max ext}}) = \text{N.A.}$$

where n_{ext} is the index of refraction of the external medium (1.0 for air) and θ_{ext} is the external angle of the incident ray relative to the fiber axis. We see that the larger the index of refraction difference and thus the larger the numerical aperture, the larger the angle a ray can have relative to the fiber axis and still be guided. For a fiber with a 1 % index step ($\Delta = 0.01$) the external maximum angle $\theta_{\text{max ext}}$ is about $8°$ (half-angle of the total cone of captured rays relative to the fiber axis). Such a fiber has a numerical aperture (N.A.) of 0.14.

From Figure 1.9 we see that high-angle rays travel longer geometrical paths per unit axial distance along the fiber than do lower-angle rays. The delay difference between the axial ray and a ray traveling at the maximum angle for total internal reflection (critical angle) is simply given by

$$\tau_{\text{axial}} = \frac{n_{\text{core}}}{c} \quad (\text{ns/km}) \qquad (1.2.4)$$

$$\tau_{\max} = \frac{n_{\text{core}}}{c} \frac{1}{\cos(\theta_{\max})} \quad (\text{ns/km})$$

$$\Delta\tau = \tau_{\max} - \tau_{\text{axial}} = \frac{n_{\text{core}}}{c}\left(\frac{n_{\text{core}}}{n_{\text{cladding}}} - 1\right) = \frac{n\Delta}{c} \quad (\text{ns/km})$$

where c is the speed of light (3×10^{-4} km/ns). For a 1 % index difference

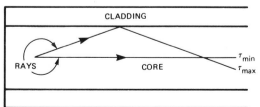

Figure 1.9. Delay spread.

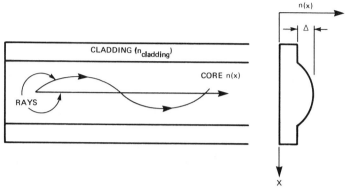

Figure 1.10. Graded index fiber.

between core and cladding, this delay difference is 50 ns/km. This delay difference between propagating rays results in a spreading in time of short duration (narrow) pulses propagating through the fiber if their energy is distributed amongst all rays. To reduce this pulse-broadening effect, one can use a graded index fiber as shown in Figure 1.10. The index of refraction tapers off slowly from a maximum value at the center of the core to a lower value at what could be defined as the core–cladding interface. Such a fiber can be made in the CVD or MCVD process by varying the composition of the layers of glass which are deposited to make the preform. Rays in a graded-index fiber propagate in helical or sinusoidal paths as shown in Figure 1.10. From a geometrical view, helical paths which wander further from the fiber axis are longer per unit axial distance. However, such helical paths "spend more of their time" in portions of the core having a lower index of refraction (which as mentioned decreases with distance from the axis). In the lower-index medium, the group velocity of light is higher. Thus rays traveling away from the axis can make up in speed what they lose in geometrical distance. With careful control of the refractive index profile, one can obtain roughly 2 orders of magnitude of reduction of the delay difference amongst rays relative to a step index fiber (having roughly the same light-ray-capturing capability). The more improvement in the delay spread (variation in group delay amongst modes) the more critical the control of the index profile is. For a fiber with a 1 % index difference between core center and cladding, a typical graded-index fiber produced in quantity might have a guaranteed delay spread of 1–5 ns/km or less (depending upon price and therefore the amount of selection in the manufacturing processes), compared to 50 ns/km for step index fibers. Because random deviations from the ideal profile, which occur during manufacture, determine the actual delay spread in any fiber, one will always be able to obtain selected ultralow-delay-spread fibers at increased cost. The boundary between standard and ultralow keeps moving downward as the technology of fiber making progresses.

There is another type of fiber, called a single-mode fiber, which is of interest for special near-term applications and which may find wide use in the future. In the above discussion we have modeled propagation in the fiber using geometrical optics (rays). As mentioned, this is a valid analytical method provided the dimensions of the fiber core are large compared to the light wavelength. One can show that there is a correspondence between light rays at certain discrete angles relative to the fiber axis, and the prop-agating electromagnetic modes of the fiber (treated as a dielectric wave-guide). Only certain rays at these discrete angles can actually propagate through the fiber. The angular spacing (relative to the axis) of these allowed rays is given by (roughly)

$$\Delta\theta = \lambda/d \qquad (1.2.5)$$

where λ is the light wavelength and d the fiber core diameter. For a fiber with a large core-diameter-to-wavelength ratio the discrete spectrum of allowed rays can be considered as a continuum, as was done above. For a core size comparable to the light wavelength, the spacing between allowed rays can be large enough that the only ray which falls below the critical angle is the axial ray itself. Thus only one ray (mode) is propagated. With only one ray, the problem of intermode delay differences disappears. How-ever, single-mode fibers are difficult to couple light into, difficult to splice, and difficult to place in cables without causing excess radiation losses due to cable bends. Thus single-mode fibers are envisioned for use in specialized near-term applications where the need for zero delay spread justifies dealing with the other problems, and also for longer-term applications where these other problems may become reduced in their impact because of technological improvements in coupling and cabling.

1.3. Sources

When lasers were first invented, communications engineers and scientists envisaged optical communication as a next step beyond millimeter waveguide systems for ultrahigh-capacity long-distance communications links. For example, one concept consisted of pipes perhaps 6 in. in diameter containing lenses or mirror pairs spaced at regular intervals carrying multiple coherent light beams. Each light beam in this "beam waveguide" could be modulated by up to thousands of telephone channels, or the equivalent in video signals, or data signals. These "light highways" would carry signals tens of miles between repeaters and hundreds of miles between telephone buildings. If a light source is shared between, say, a thousand telephone conversations (circuits), and if its output travels tens of miles before detection, amplification, and regeneration with a new source, then it is feasible to spend thousands of dollars for the source, while still keeping the prorated cost well below $1 per circuit mile.

The concept of fiber optics is aimed at shorter-distance lower-capacity links as a replacement for wire pairs. Because of the smaller number of circuits sharing the source, and the shorter distance of propagation between sources, the amount of money that can be spent on the source is reduced to a few tens or hundreds of dollars. Further, to be compatible with the fiber dimensions and to provide high conversion efficiency of electrical input power into guided light output power, the possible candidates for fiber systems are limited to a few inexpensive, efficient, easy to modulate solid state devices.

Basic Source Types for Fiber Systems and Parameter Definitions

The leading contenders for fiber optic sources are the light-emitting diode (LED) and the injection laser. Both of these devices are microscopic in size, relatively efficient in their ability to convert electrical power to useful light output power, potentially inexpensive in large-quantity production, relatively easy to modulate, and potentially reliable with careful fabrication techniques and structural design.

Present readily available LEDs and lasers for fiber applications are fabricated from gallium arsenide and gallium aluminum arsenide with up to 30% aluminum. These devices emit in the 0.75–0.9-μm wavelength range. Devices fabricated from gallium indium arsenide phosphide are available but are not as highly developed. These GaInAsP devices emit in the 1–1.5-μm wavelength range where fiber losses are lower.

The simplest form of light-emitting diode consists of a forward-biased *pn* junction fabricated from a semiconductor whose band gap energy corresponds to the energy of a photon at the desired emission wavelength (see Figure 1.11). The material must be a direct band-gap material, which means that holes and electrons injected into the junction tend to combine directly to emit photons rather than interactively with the lattice to emit phonons (heat in the form of lattice vibrations). Doping levels must be chosen so that the lifetime of carriers in the junction is small. This insures that the carrier density, and therefore the emitted light output, will respond quickly to variations in the injected current. Thus the device light output can be directly modulated by varying the current rather than requiring some external light modulator. The current must be confined to a narrow column under the fiber core (for example, by using an oxide layer to mask off and define a small dot contact) so that as much of the generated light

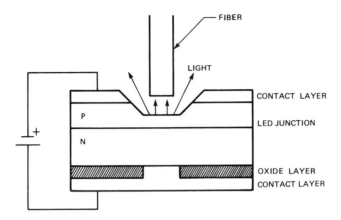

Figure 1.11. Simple LED (homojunction).

as possible can be captured. To avoid reabsorption of the emitted light a well can be etched over the recombination region to bring the fiber into close proximity to the junction.

A problem with this structure is that it is difficult to confine the injected carriers to a thin layer near the junction and to bring the fiber sufficiently close to the junction to avoid reabsorption. If the well is etched too deep, carriers will combine nonradiatively at the surface of the well owing to surface defects and impurities. To avoid this problem one can use the multilayer heterostructure approach shown in Figure 1.12. In this approach, materials having different energy level structures but with reasonably closely matched crystal lattice parameters are grown in sequential layers to form energy barriers that tend to confine holes and electrons to the junction region. In addition, light emitted by holes and electrons combining near the junction in the "active layer" is only weakly absorbed by the materials surrounding the junction. Thus the etched well need not come so close to the junction as to make surface recombination a problem. Another feature of the structure shown in Figure 1.12 is that the device is mounted substrate up on the heat sink so as to bring the recombination region in close thermal contact to the heat sink.

Since the light-emitting diode is operated at finite temperature there is a spread in available energy levels in the conduction and valence bands given by approximately kT, where k is Boltzmann's constant and T is absolute temperature. This results in a fractional bandwidth of the emitted light of approximately $kT/h\Omega$, where $h\Omega$ is the energy in a photon (nominal band-gap energy). In typical fiber applications this fractional bandwidth is about 0.05.

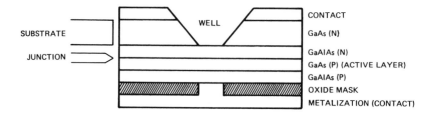

Figure 1.12. Heterojunction LED.

The light emitted by the light-emitting diode is incoherent (noiselike) in nature. The above fractional bandwidth corresponds to an actual spectral width of about 10^{13} Hz, far in excess of the bandwidth of any modulating signal. The LED output can be characterized as a Gaussian random process which implies, among other things, that its amplitude is a Gaussian random variable whose value randomizes about every 10^{-13} s.

The rate at which a light-emitting diode output can respond to changes in the current injected into the device is governed by the recombination lifetimes of carriers in the junction. If the injection current is turned off, the population of carriers will exponentially decay as will the emitted light. Typical decay times for gallium arsenide devices, which emit in the 0.8–0.9-μm region, are on the order of a few nanoseconds. The exact decay time constant depends upon the doping levels in the device. There is a tradeoff between response time and output power.

Figure 1.13 shows the light output into a typical transmission fiber

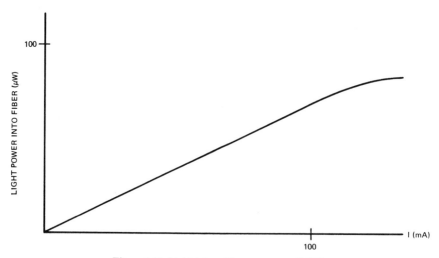

Figure 1.13. Light into a fiber vs. current (LED).

vs. applied current for a typical high-brightness light-emitting diode. The characteristic is fairly linear up to the point where saturation due to heating effects begins. As shown, currents are typically in the 100-mA range for maximum output.

The rate at which a diode can actually be modulated depends upon the recombination lifetime as mentioned above, and also upon the circuit providing the current drive. LED capacitances are on the order of tens of picofarads, and high-current low-impedance drivers are required for maximum-modulation bandwidths. Modulation of light-emitting diodes at rates up to 20 Mb/s (Megabit per second) is considered routine. Beyond 100 Mb/s carefully selected devices and sophisticated driver circuits are usually required.

The other source of interest for optical fiber applications is the injection laser. Figure 1.14 shows a typical gallium aluminum arsenide device (other material systems, e.g., GaInAsP, have been used to make diode lasers, but GaAlAs is presently the most highly developed).

The structure consists of a sequence of layers of varying aluminum content and with various dopants to make the layers p or n type. The active region at the junction is grown with between 0 and 10% aluminum depending upon the desired emission wavelength. Pure GaAs has a band gap resulting in about a 900-nm emission, while 10% AlGaAs has an ~800-nm emission wavelength. The active region (where holes and electrons combine) is sandwiched between layers of AlGaAs with about 30% aluminum. This results in a potential well that confines holes and electrons to the active region. In addition, since the index of refraction of the 30% AlGaAs is

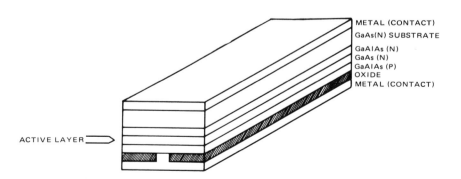

Figure 1.14. Gallium arsenide injection laser.

lower than the active region index, light generated in the active region is confined in the direction perpendicular to the junction by waveguiding action.

The cleaved faces of the laser diode act as partially reflecting mirrors since the reflection coefficient between GaAs and air is around 30%.

Light confinement in the direction parallel to the junction results from variations of the index of refraction of the active region with carrier injection. That is, light tends to be confined to the portion of the junction that carries the injected current. This portion is defined by one of a number of techniques. Figure 1.14 shows the oxide mask approach.

The cavity defined by the three directions of confinement has a number of resonant modes with associated emission frequencies. Typical laser diodes will oscillate in a number of modes simultaneously, and in addition can switch between groups of modes in a random fashion. In many applications it is desirable to fabricate a laser in such a way as to eliminate this mode hopping, and even to obtain single-mode (frequency) operation. This is accomplished by careful selection of the device geometry and composition.

Even when multimode oscillation and mode hopping occur, the spectral width of the laser emission is much narrower than the light-emitting diode. Typical spectral widths are less than 1 nm for the laser compared to 40–50 nm for an LED. The narrow spectral width becomes important when material dispersion (pulse spreading in transmission due to variation of the group velocity of light in glass with wavelength) is a factor.

Even if the laser oscillates in a few modes simultaneously more than 50% of its output can be coupled into a multimode fiber, compared to a few percent or less coupling efficiency for an LED. Thus a laser emitting the same total output power as an LED can couple 15–20 dB more light into a fiber.

The light output vs. current characteristic of a laser is shown in Figure 1.15. The threshold of rapid increase in light vs. current occurs at the point where the carrier density in the device is sufficiently great to have stimulated emission exceeding reabsorption of light and radiation. The laser is generally biased near or above threshold for high-speed modulation in order to avoid the delays associated with increasing the carrier density to the threshold level. The device is easily damaged by currents which are too far above threshold, generally resulting in facet (emission face surface) damage from excessive field intensities. Since the threshold current depends upon the device temperature and can also vary with device age, driver circuits for lasers usually include complex bias control circuitry. This will be discussed in detail in the chapter on transmitters. When properly biased, lasers are capable of being directly modulated at rates of beyond 1 GHz. This improvement over the LED characteristics is due to the faster time constant

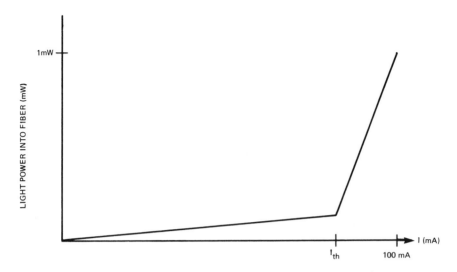

Figure 1.15. Light output into a fiber vs. current (injection laser).

of stimulated emission vs. spontaneous recombination and to the smaller incremental currents needed to modulate the laser around its bias point.

1.4. Detectors[†]

The function of the detector in a fiber optic communication link is to convert optical power into an electrical response. The most common detector example in fiber applications is the photodiode, which acts as a converter of optical power to electrical current. Basically detectors can be divided into two types: instantaneous photon-to-electron converters and thermal parametric types. The photodiode is of the former variety, where absorbed photons generate hole–electron pairs to produce an electrical current. An example of the latter type would be a bolometer, where absorbed light energy causes the resistivity of the device to vary. For high-speed applications, instantaneous photon-to-electron converters are generally required.[‡]

[†] See Chapter 18 of Reference 4 and Chapter 3 of Reference 5.

[‡] It is not necessary to quantize the optical electromagnetic field into photons in order to describe the operation of photodiodes; however, for the purposes of this book, we shall do so.

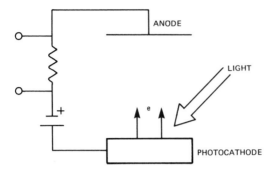

Figure 1.16. Vacuum photodiode.

1.4.1. Photodiodes: Physical Description

The simplest example of a photon-to-electron converter is a vacuum photodiode as shown in Figure 1.16. It consists of a photocathode made of a photoemissive material, and an anode, all enclosed in a vacuum container. If the top surface of the photocathode is illuminated, electrons in the photocathode will absorb energy from the light in quantized units or photons. The energy in a photon is given by $E = hf$, where h is Planck's constant and f is the optical frequency. If this energy is sufficiently large, atoms in the photoemissive surface can be ionized, and the resulting liberated electrons can be emitted from the photocathode surface. The electrons can also lose all or part of their absorbed energy in the process of traveling toward the photocathode surface, and as a result have insufficient energy to overcome the surface potential. The fraction of absorbed photons which result in liberated electrons is called the quantum efficiency of the photocathode. Since the optical power falling on the photocathode is hf multiplied by the number of photons incident per second, and since the emitted current is the electron charge e multiplied by the number of electrons liberated per second, it follows that the "responsivity" of the photocathode in amps of current per watt of incident light is related to the quantum efficiency by the equation

$$R = \eta e / hf \qquad (1.4.1)$$

where η is the quantum efficiency (dimensionless), R is the responsivity in A/W, e is the electron charge, 1.6×10^{-19} C, and hf is the energy in a photon (joules).

The responsivity and the quantum efficiency are alternative ways of describing the efficiency with which light is converted to current. Figure 1.17 shows some curves of these parameters of typical photocathodes.

When an electron is emitted by the photocathode, it travels toward the anode under the influence of the electric field within the photodiode.

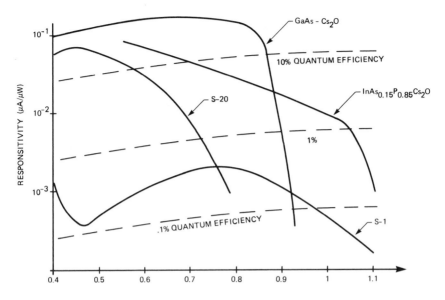

Figure 1.17. Responsivity of typical photocathodes.

This results in a displacement current flowing through the load as shown in Figure 1.16. The duration of the response associated with a single liberated electron is proportional to the transit time from cathode to anode. The duration of the photodiode response to a short pulse of light liberating many electrons is therefore also proportional to this transit time. Thus for a high-speed detector, this transit time must be as short as possible. Quantum efficiency (or responsivity) and response speed are the two important performance characteristics of photodiodes.

The solid state equivalent of a vacuum photodiode is shown in its most simple form in Figure 1.18. If a pn junction is back biased as shown, mobile holes and electrons move away from the junction leaving behind changed acceptor and donor atoms. From Gauss's law one can calculate the resulting electric field and in turn the voltage across the junction. The results are (6)

$$\rho_n W_n = \rho_p W_p$$
$$E_{max} = \rho_n W_n / \varepsilon = \rho_p W_p / \varepsilon$$
$$V = (\rho_n W_n^2 + \rho_p W_p^2)/2\varepsilon \qquad (1.4.2)$$

where ρ_n is the doping level in the n material, ρ_p is the doping level in the p material, W_n is the width of the depletion region (see Figure 1.19) on the n side, W_p is the width of the depletion region on the p side, V is the voltage

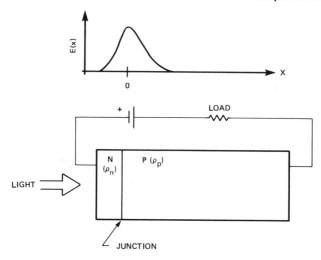

Figure 1.18. Simple *pn* diode.

across the junction, E_{max} is the maximum field, and ε is the material permittivity.

If we assume $\rho_p \ll \rho_n$, then

$$W_p = (2\varepsilon V/\rho_p)^{1/2}$$
$$E_{max} = (2V\rho_p/\varepsilon)^{1/2} \tag{1.4.3}$$

When the photodiode is illuminated as shown in Figure 1.19 photons of light can be absorbed to generate hole–electron pairs. If these hole–electron pairs are generated in a region of significant electric field (depletion region) then they will move at saturation-limited velocities to produce a large displacement current (displacement current is proportional to the product of carrier velocity and local electric field level). If hole–electron pairs are generated outside of the depletion region, then the electron will randomly and slowly diffuse into the depletion region to produce a delayed displacement current.

The amount of penetration of the incident light into the photodiode depends on the material and the light wavelength. To make a fast device, one tries to arrange for the depletion region to be at least as wide as the light absorption (penetration) region. From equation (1.4.3) we see that a wide depletion region is obtained by reducing the doping level ρ_p. If this doping level is sufficiently low the p region becomes essentially intrinsic (*i* type). In order to make a good ohmic contact on the right, a heavily doped p region must be added. One then ends up with the pin (right to left) structure shown in Figure 1.20. Figure 1.21 shows curves of quantum efficiency and responsivity for typical high-speed photodiodes.

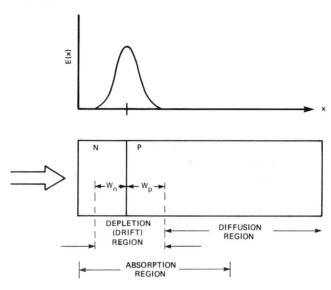

Figure 1.19. *pn* diode response.

In designing the solid state photodiode there is a tradeoff between quantum efficiency and speed of response. For high quantum efficiency, the width of the device (*i* region) should be comparable to two or three times the absorption length of light in the material being used. For high speed of response the device should be as thin as possible to minimize carrier transit times. For silicon detectors operating at wavelengths below

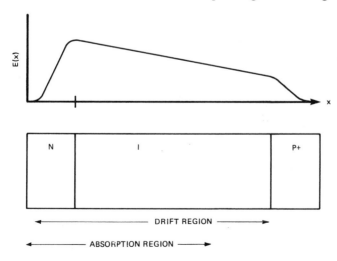

Figure 1.20. Pin diode.

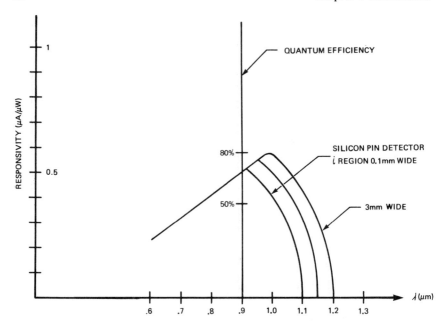

Figure 1.21. Photodiode responsivity.

0.9 μm, high quantum efficiency (in excess of 50%) and fast response (less than 0.5 ns) are simultaneously obtainable. At longer wavelengths, as the absorption length in silicon rapidly increases, other materials are required.

An important parameter in solid state photodiodes in addition to quantum efficiency and speed of response is dark current. Dark current refers to the leakage in a back-biased diode in the absence of illumination and is caused by thermal generation of hole–electron pairs. The amount of dark current is related to the temperature of the device, the band gap of the material, and the care taken to control surface leakage. Silicon devices can be fabricated with extremely low dark currents. Materials used at longer wavelengths (e.g., germanium and quaternary compounds) have significantly poorer dark current control.

1.4.2. Avalanche Photodiodes[7-10]

As we shall see in the chapter on receiver design (Chapter 3), sensitive optical receivers operate at minimum detectable power levels of the order of a nanowatt. At typical fiber optic system wavelengths, even a high-quantum-efficiency photodiode would produce a response to this power level on the order of a nanoampere. Such weak currents are severely corrupted by noise from electronic amplifier stages following the detector. In order to obtain acceptable receiver performance, a means is needed to increase the detector output before electronic amplification. This can be accomplished with an avalanche photodiode as shown in Figure 1.22. Photoelectrons created by light absorption in the absorption region drift or diffuse into the "high-field" region where saturation-limited velocities are very high. Occasionally between collisions with the lattice one of these carriers will develop sufficient energy to cause an ionizing collision producing an additional hole–electron pair. New carriers generated in this fashion can also produce additional carriers by collision ionization, etc. Thus a single "primary" hole–electron pair generated by incident light can produce tens, hundreds, or more "secondary" pairs by collision ionization. The resulting photocurrent is correspondingly tens or hundreds of times greater.

The avalanche multiplication process is statistical. A primary photo-electron produces a random (unpredictable) number of secondaries, governed by a complex statistical description to be discussed in Chapter 3. The average number of secondary electrons produced by a primary electron is dependent upon the material, the electric field level in the high-field region, and the width of the high-field region. Thus the average gain of the device is dependent upon the device geometry and the reverse bias applied. The statistics of the multiplication process are also dependent

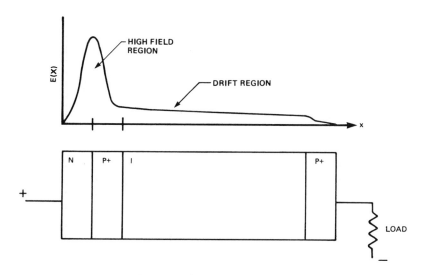

Figure 1.22. Avalanche photodiode.

upon the geometry, as well as upon the material, the material uniformity, and the applied voltage.

The mean avalanche gain is dependent upon the temperature of the device because the mean free path between nonionizing collisions with the lattice decreases as the device temperature increases. Thus at high temperatures the probability of a carrier acquiring sufficient energy between collisions for an ionizing collision decreases. Figure 1.23 shows a typical curve of average gain vs. applied reverse bias and temperature for a silicon avalanche photodiode. The steepness of the curves at high average gains, and the temperature sensitivity, make avalanche diodes more difficult to use than pin detectors, as will be discussed in Chapter 3.

Referring to Figure 1.22, we see that to make a fast device it is necessary to have a lightly doped (intrinsic) drift region to the right of the heavily doped high-field region, which can be depleted of carriers (swept out). In this depleted drift region optically generated primary photoelectrons can move rapidly under the influence of the electric field toward the high-field region to produce a fast response. As the back-bias voltage is increased from zero, first the high-field region is fully depleted, and then the drift

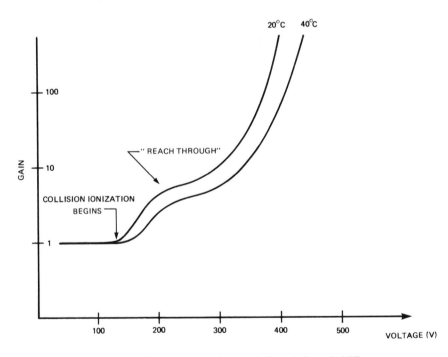

Figure 1.23. Gain vs. voltage for a typical reach through APD.

region is depleted (swept out). If the doping levels and geometry are chosen so that the high-field region is fully depleted at field levels which are below the values required for avalanche multiplication, then very high back-bias voltages will be required to produce the further field level increases in the high-field region needed to obtain multiplication. If, on the other hand, the field levels in the high-field region reach values sufficient to produce multiplication much before the high-field region is fully depleted, then at desired values of gain the drift region may not be swept out, resulting in a slow-speed device. Careful tailoring of the doping profiles and precise fabrication control are needed to produce a device which has adequate gain at reasonable reverse-bias voltages but which can be operated at reduced gain with reduced reverse bias and still retain a fast speed of response.

Typical silicon avalanche photodiodes can be fabricated to operate at back-bias voltages near 300 V with gain ranges adjustable from less than 10 to over 100, and having response speeds of less than 1 ns.

Problems

1. A step index fiber has a core index of refraction of 1.5 and an index step Δ of 0.02. Calculate the maximum angle which a guided ray may have relative to the axis inside the fiber; outside the fiber. Calculate the maximum delay difference per unit length between the axial and a guided nonaxial ray. For the above use both "exact" and approximate relationships, and compare the results [see equation (1.2.1)].

2. Consider a step index fiber with a core index of refraction of 1.5, and an index step Δ of 0.01. How small must the core be so that only a single mode is guided (assume $\lambda = 0.9$ μm and the angular separation between modes is approximately λ/D)?

3. Calculate the center frequency, f, of a light source emitting at $\lambda = 0.9$-μm wavelength. Calculate kT/hf at $T = 300$ K. Calculate the bandwidth of an LED which is approximately $(kT/hf) \cdot f$.

4. Solve equation (1.4.2) for W_p as a function of V, ρ_n, and ρ_p. For $V = 25$ V and $W_p = 1$ mm, calculate the required immobile charge density ρ_p (assume $\rho_n \gg \rho_p$). Convert this to a doping level in donor atoms per cm^3. (Assume a silicon device.)

5. Suppose a digital pulse transmitter emits 1 mW of average light power, and a corresponding digital receiver requires an average optical signal input energy of 10^{-16} J per received pulse (for satisfactory reception). Suppose a fiber has a loss of 10 dB km^{-1}. Calculate and plot the maximum fiber length between the transmitter and the receiver vs. the pulse rate, for rates between 10^5 and 10^9 pulses per second. (Ignore coupling losses into and out of the fiber.)

6. In the above example, assume that, in addition, pulses traveling through the fiber spread in time due to delay distortion by 5 ns km^{-1}. If the maximum allowable spreading must not exceed 0.5/pulse rate, calculate and plot the maximum allowable fiber length vs. the pulse rate.

Requirements for Systems and Subsystems [1, 2]

2.1. Requirements

Before going into the details of transmitter, receiver, and system design in Chapter 3 it is necessary to define the requirements which lead to those designs. In particular we must first describe the types of signals to be transmitted by a fiber link (or any other link); and we must discuss the practical constraints such as powering limitations, environmental requirements, and dynamic range requirements which affect details of practical designs. We shall see that there are a number of major applications which are different enough so that significant differences exist in the fiber subsystems needed to meet these applications' requirements. In the sections below we shall begin with a review of the major categories of signals to be transmitted over fiber links, and the requirements on performance (fidelity) in transmission. This will be done mostly from an analytical or "systems engineering" viewpoint. Then we shall examine some practical constraints on powering, etc., as mentioned above.

2.2. Signals [3, 4, 5]

From an analytical viewpoint we can facilitate the discussion of signals by dividing them into four categories: clocked binary signals, unclocked binary signals, multilevel signals, and analog signals. Figure 2.1 shows a typical application where a signal source producing one of these signal types, in the form of a voltage waveform, drives a transmission link, which

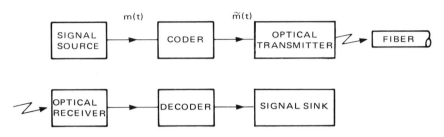

Figure 2.1. A typical system.

carries the waveform to a distant signal sink. The performance of the transmission link is based on the similarity between the input and output waveforms. In each application this similarity is quantifiable via a defined fidelity criterion as will be discussed below. Before we can proceed to design fiber systems and subsystems we must understand what these fidelity criteria are and what factors in the design affect these measures of performance.

2.2.1. Clocked Digital Systems

In many important applications the signal source (see Figure 2.1) generates a binary signal (two defined voltage states) which can make transitions from one state (voltage) to the other only at well-defined periodically spaced points in time. We can call this the message signal. A typical waveform is shown in Figure 2.2. Very often the signal source provides a second periodic, deterministic, signal output called a "clock," which is synchronized with the message signal as shown. We shall assume in what follows that the clock signal is indeed available to the transmission system at the input. Later in Section 3.4.6 on receiver design we shall show how a clock signal can be synthesized locally from the message signal if necessary.

The rate at which the clock runs is called the "baud" or symbol rate (baud, not baud rate). In each "time slot" a pulse may or may not be present. The width of a pulse in a time slot compared to the time slot width is called the duty cycle. If the duty cycle is less than one then we have a "return-to-zero" (RZ) signal. That is, even if we have a continuous sequence of pulses, the signal returns to zero level in each time slot. If the duty cycle is one, then conversely we have a "non-return-to-zero" (NRZ) signal.

The fidelity requirement is usually related to "errors." That is, in each time slot there is a chance that the output signal will contain a pulse while the input signal did not (false alarm) or vice versa (miss). Errors are measured in a number of standard ways. One approach is to take the ratio of the average number of errors per unit time to the number of pulses transmitted per unit

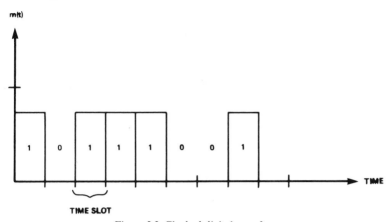

Figure 2.2. Clocked digital waveform.

time. This is called the "error rate," expressed as a number such as 10^{-9} (one error in a billion pulses on the average). Sometimes errors occur in bunches or bursts and the performance of the system is described in terms of "error-free seconds" or "errored seconds." That is, if one measures errors which accumulate in 1-s intervals, then one can talk of, say, "one errored second per day," meaning that in 24 hr only one of the 1-s intervals (on the average) contains any errors.

Note that a distinguishing feature of a binary clocked signal is that for the most part the transmission fidelity is not governed by the details of how the signals may be distorted in transmission, but only by the errors. This is because the waveforms can be reshaped and reclocked by nonlinear means at the receiver output using standard "regenerator" circuits. There is, however, one other consideration: This is called jitter. It happens that, in the process of transmission, the periodicity of the spacing between pulses can be disturbed by a number of effects to be discussed later in Section 3.4.6. Jitter is shown (exaggerated) in Figure 2.3. In principle jitter can be removed

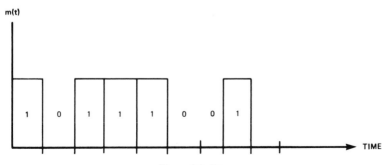

Figure 2.3. Jitter.

from the output signal in many cases using various clock-averaging and reclocking techniques incorporating "elastic stores." However, in general, it is desirable to avoid jitter accumulation and therefore the necessity for such jitter-removing devices.

In order to design a transmission system for binary clocked signals, certain statistics of the signals must be known. Obviously the clock rate or baud is of major importance. Another major concern relates to the effects of ac coupling internal to the transmission link.

To understand the problem consider Figure 2.4. Here an isolated pulse from a clocked binary signal enters a filter which does not pass dc. The output of the filter is a pulse with a "tail" of opposite polarity. It is easy to show that the output pulse must have zero area by definition of the ac coupling of the filter:

$$H_{\text{filter}}(f)|_{f=0} = 0 \qquad \text{(by definition)} \tag{2.2.1}$$

$$H_{\text{out}}(f) = H_{\text{in}}(f)\, H_{\text{filter}}(f)$$

$$\text{area}\left[h_{\text{out}}(t)\right] = \int h_{\text{out}}(t)\, e^{i2\pi f t}\, dt\Big|_{f=0} = H_{\text{out}}(f)\Big|_{f=0} = 0$$

where $h_x(t)$ is function of time (t) and where $H_x(f)$ is the Fourier transform of $h_x(t)$ as a function of frequency f.

The amplitude and duration of the tail depend upon the exact low-frequency characteristics of the filter. In general, the lower the low-frequency cutoff of the ac-coupling filter, the longer and lower is the tail.

Now imagine a sequence of pulses passing through the ac coupling

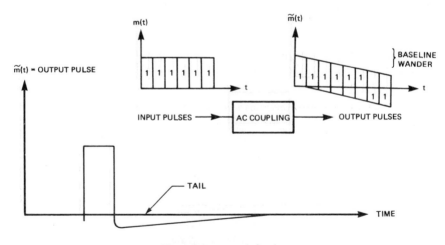

Figure 2.4. ac-coupled pulses.

filter as shown in Figure 2.4. The tails of individual pulses can accumulate
to cause the "baseline wander" effect, shown exaggerated in the figure. If
baseline wander occurs internally in a fiber transmission link, the ability
of the receiver circuitry to distinguish between pulses present and pulses
absent can be hampered. Thus errors can be caused, or made more likely
to occur. Various means can be devised to limit baseline wander.

One approach is to make use of the coder shown in Figure 2.1. The
function of the coder is to transform the input waveform into a form more
suitable for transmission. The decoder reverses this transformation. An
example of a coding scheme that alleviates baseline wander is bipolar coding,
as shown in Figure 2.5. The bipolar coder converts each incoming pulse

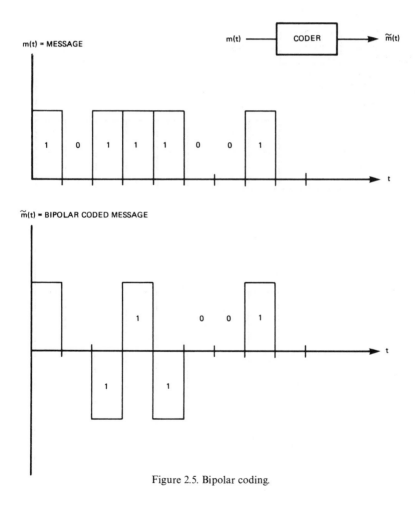

Figure 2.5. Bipolar coding.

alternately into a positive or negative pulse. Thus the two-level (binary) input signal is converted into a three-level (ternary) output signal. This operation is easily reversible at the decoder, since positive and negative pulses in the ternary signal both represent positive pulses in the binary signal. The effect of bipolar coding is to balance the signal with respect to dc. That is, the tails of ac-coupled individual pulses will tend to cancel because of the alternating polarity of individual pulses. The price one pays for this is the necessity of transmitting a ternary signal. The scheme is inefficient in the sense that a true ternary signal can carry $\log_2(3) = 1.58$ bits of information per symbol, whereas the binary information contains only 1 bit per symbol.

Another coding scheme is "Manchester" coding (and its variations). Here each binary pulse is coded as shown in Figure 2.6. An input binary "space" causes a change in level at the coder output at the beginning of a time slot. An input binary "mark" results in an output transition at the beginning and the middle of a time slot. In effect, using this scheme the symbol rate is doubled since the transmission link must resolve a coded time slot of half the width of the time slot of the signal from the source. If a Manchester-coded signal is ac coupled, only an inconsequential fixed dc offset occurs, as shown in Figure 2.6. This is due to the balance in the

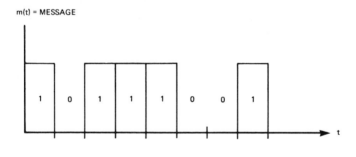

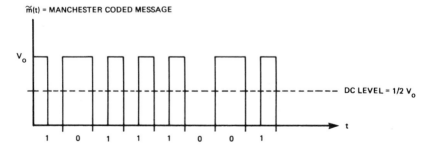

Figure 2.6. Manchester coding.

density of pulses and spaces imposed by the code. Since the transmission link must transmit essentially at twice the symbol rate of the uncoded signal, we again trade off transmission efficiency against the ability to tolerate ac coupling.

There are more efficient coding schemes which convert several binary pulses (taken as a "block") into blocks of fewer ternary pulses or more binary pulses, resulting in a balance between pulses of opposite polarity or in a balance in the density of pulses and spaces. These are discussed in detail in the literature. The tradeoff is in the complexity of the coder–decoder pair vs. efficiency in the use of the transmission link.

Another approach to alleviating baseline wander is to use "clamping" in the receiver. Clamping is illustrated in Figure 2.7. If the filter output pulses return to zero in each time slot, then the nonlinearity of the diode can be used to "clamp" the baseline to zero. This scheme requires adequate bandwidth internal to the fiber link to allow the return to zero of the ac-coupled pulse stream. Thus, it too, is somewhat inefficient.

A final method of controlling baseline wander to be discussed here is "scrambling." The scrambler is a device which attempts to limit the unbalances in the numbers of pulses and spaces which can occur in typical binary signals from standard signal sources. In effect the scrambler attempts to randomize the incoming signal pattern using a reversible coding operation. To the extent that the scrambler can do this, the effects of the scrambling operation are to guarantee a predictable dc offset due to ac coupling, and a predictable (statistically) wander from the offset, which can be taken account of in the system design.

On the average the pulse sequence at the ideal scrambler output would be one-half marks and one-half spaces (ones and zeros). The tails would

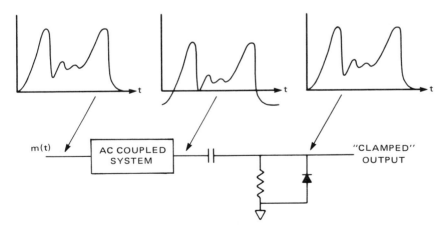

Figure 2.7. Clamping.

accumulate to produce an inconsequential dc level plus a random variation from this level. For computational purposes it is often assumed that the sum of the tails is a Gaussian random variable, because typically so many tails contribute to this sum that the central limit theory should apply. The baseline wander at any point in time is therefore completely characterized by its mean and variance. These quantities are obtained as follows: Assume the tail of the ac-coupled pulse is exponential in shape with an area equal to the area of a filter input pulse and with a decay time constant N times as long as a time slot T. Let the peak value of an input pulse be given by (see Figure 2.8) $h_{in}(t)|_{peak} = [H_{in}(0)/T] K$, where K is a "peaking factor." Then for a random input sequence to the ac-coupling filter we obtain

$$\text{average offset} = \left\langle -\sum_{k=0}^{\infty} a_k [H_{in}(0)/NT] e^{-(t-kT)/NT} \right\rangle \qquad (2.2.2)$$

where $a_k = 0$ or 1 with equal probability $= 1/2$, the angle brackets indicate average value, and $H_{in}(0)$ is the area of an isolated input pulse. Therefore

$$\text{average (dc) offset} = -\tfrac{1}{2} H_{in}(0)/T$$

$$\text{offset variance} = \left\langle \sum_{k=0}^{\infty} \sum_{j=0}^{\infty} a_k a_j [H_{in}(0)/NT]^2 \right.$$
$$\left. \times e^{-(t-kT)/NT} e^{-(t-jT)/NT} \right\rangle - (\text{average offset})^2$$
$$= \tfrac{1}{8} H_{in}^2(0)/NT^2$$

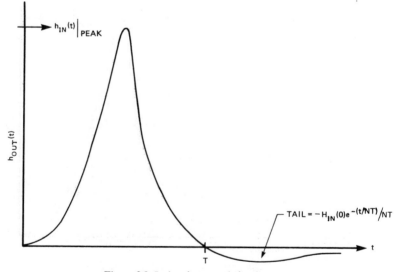

Figure 2.8. Isolated ac-coupled pulse.

Thus we obtain

$$\frac{[h_{in}(t)/\text{peak}]^2}{\text{offset variance}} = 8K^2N \qquad (2.2.3)$$

Thus we see that the longer the tail is compared to a time slot (i.e., the larger N is) the smaller is the baseline wander standard deviation compared to the peak of an isolated pulse. This ratio improves in proportion to the square root of the quantity N. If we wish the baseline wander standard deviation to be, say, 3 % of the peak pulse height, and if K is, say, 1 (corresponding, for example, to a full duty cycle pulse), then from equation (2.2.3) we must have N equal to approximately 125. Since NT is the decay constant of the tail this means that the low-frequency cutoff of the ac-coupling filter must be less than $B/(2\pi \cdot 125)$ or about $B/800$, where B is the symbol rate $1/T$.

The standard deviation and mean of the baseline wander characterize it to the extent that it is approximated by a Gaussian random variable. More exact statistical calculations may be necessary in some circumstances to be discussed later in Section 3.4.4 on receiver design.

In order to obtain a random data sequence, some sort of scrambler as mentioned above is required. One popular practical scrambler–descrambler combination is called the "self-synchronizing" type as shown in Figure 2.9. One can show that the tandem operations of scrambling and descrambling reproduce the original input sequence, as follows:

$$y(t) = m(t) \oplus \sum_{\substack{\text{taps} \\ \text{used}}} y(t - kT)$$

therefore

$$y(t) \oplus \sum_{\substack{\text{taps} \\ \text{used}}} y(t - kT) = m(t)$$

$$x(t) = y(t) \oplus \sum_{\substack{\text{taps} \\ \text{used}}} y(t - kT)$$

therefore

$$x(t) = m(t)$$

Certain combinations of taps and scrambler register lengths L give maximum scrambling, in the sense that for an all-zero input the scrambler (once initialized with some nonzero state) will produce all possible binary sequences of length L, except all zeros ($2^L - 1$ sequences). Table 2.1 shows

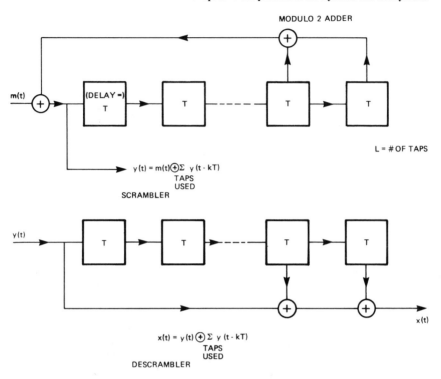

Figure 2.9. Self-synchronizing scrambler–descrambler.

typical "maximal length" scramblers having this property. Unfortunately, self-synchronizing scramblers of the type described above do not produce a truly random sequence. In particular, imbalances in the numbers of ones vs. zeros can accumulate for periods of time sufficient to cause baseline wander more often than one might expect from a true random sequence. The longer the scrambler length, the more serious the problem. Thus although the scrambler is very useful for breaking up long sequences of ones and zeros in the input data stream, the scrambler register length, L, should be limited to less than about 17 in order to avoid excessive low-frequency content in the scrambled sequence. The calculation of the low-frequency content is extremely difficult, and typically brute force computer simulation is used to obtain numerical results. One finds that for scramblers of length L less than or equal to 17, to reduce baseline wander to a few percent of the peak pulse height a low-frequency ac-coupling cutoff of less than about 0.01% of the symbol rate B is required. For example, a 50-Mbaud system would require a 5-kHz or lower cutoff.

**Table 2.1. Scrambler
Tap Locations**

Length	Taps used
5	3, 5
7	4, 7
9	5, 9
11	9, 11
15	14, 15
17	14, 17

2.2.2. Unclocked Signals

For many data bus applications the signal impinging upon the transmitter subsystem is a binary signal with well-defined levels but arbitrary transition times. The signal can remain in the high or low state for arbitrarily long periods of time. The information is contained not only in the state but in the time that a transition has taken place. The performance of the optical link is typically governed by how well the transitions can be reproduced at the receiver subsystem output both in rise time and in position. Typically rise-time limitations and jitter in the transition times are degradations that accumulate through the link and are difficult to remove.

Several approaches have been used in optical fiber systems to interface with unclocked signals. One straightforward approach is to oversample the incoming signal as shown in Figure 2.10. This converts the unclocked signal to a clocked signal which can be dealt with as described in Section 2.2.1 above. Sampling introduces an uncertainty in the transition time equal to the sample spacing. The requirements on the allowable jitter determine the sampling rate and thus the required bandwidth of the fiber link.

Another technique is shown in Figure 2.11. In this coding scheme positive-going transitions at the coder input produce a positive pulse at the coder output and negative-going input transitions produce a negative output pulse. The three-level signal can be ac coupled since positive and negative pulses must alternate. The original signal is recovered at the receive end with a decoder which uses two thresholds to convert positive pulses to positive transitions and negative pulses to negative transitions as shown in Figure 2.11. The jitter added in this coding and decoding operation depends upon the rise times and signal-to-noise ratios at the encoder and decoder. This will be quantified in Section 4.3.3 on pulse position modulation systems.

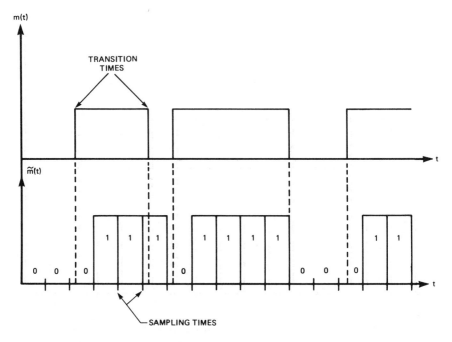

Figure 2.10. Oversampling an unclocked signal.

2.2.3. Multilevel Signals

Many of the requirements for clocked and unclocked multilevel signals are analogous to their binary counterparts. However, multilevel signal transmission results in additional restraints on linearity and gain control in the transmission link. In binary applications one must simply decide at the receiver whether a signal is above or below a threshold. In multilevel operation, one must ask how big the signal is compared to some reference. Therefore, as mentioned, requirements an automatic gain control and linearity become more severe.

2.2.4. Analog Signals

Analog signals take on a continuum of levels. Typically, but not always, they can be assumed to be amenable to ac coupling. However, the requirements in the fidelity of the transmission are generally more severe than for binary or multilevel transmission. A typical measure of transmission quality is the "signal-to-noise ratio." A typical analog system is shown in Figure 2.12. One assumes that the demodulator output signal is the sum of the

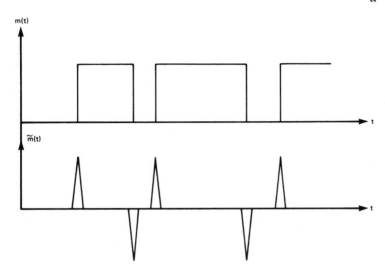

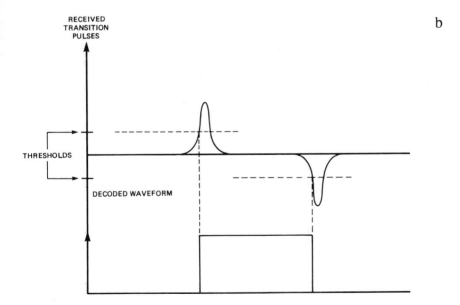

Figure 2.11. (a) Three-level transition coder. (b) Three-level transition decoder.

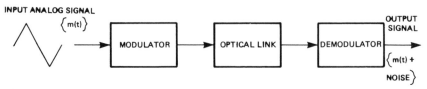

Figure 2.12. Typical analog system.

modulator input signal and an "error" signal. This error signal is caused by noise and nonlinearities in the link. The performance measure is the ratio of the rms value of the error signal to the peak or rms value of the message signal. Ratios of 50–70 dB are common requirements. Furthermore, in analog links these degradations tend to accumulate as repeaters are placed in tandem. To alleviate the requirements on linearity and noise, various modulation schemes can be employed.

One approach is to convert the analog signal to digital form using a coder. For example, a 3-kHz nominal bandwidth voice signal can be converted to 64-kb/s binary signal using standard hardware. Video signals with 4-MHz bandwidth are typically coded to 90 Mb/s using existing hardware, although lower digital rates for video can be obtained with sophisticated coding techniques.

Various forms of frequency modulation or sampling followed by pulse position modulation can be used to make the analog signal more immune to the degradations which will occur in transmission over the fiber link (6, 7). These will be treated again in Section 4.5.3 on analog systems.

2.3. Practical Constraints

We shall now address some practical issues which affect system and subsystem design. These include powering, environmental requirements, and operational requirements such as dynamic range.

2.3.1. Powering for Subsystems

When designing a subsystem for use in applications outside of the laboratory, it is important to recognize the restrictions which typically exist on powering. Except for instances where very small current drains are required, each dc voltage requires a separate power supply in the equipment rack, including wiring, fusing, automatic backup (for protected systems) and associated maintenance. Thus it is prudent to limit the number of voltages required to as few as possible, selecting values which are likely to be common to other equipment in the system. For example, equipment employing TTL logic circuitry will have +5 V available, while equipment employing ECL logic circuits will have − 5.2 V available.

Power consumption is important because of the constraints placed on heat removal, power supply size, and energy usage (particularly in remote systems powered by a battery or from distant supplies).

All real subsystems must allow for sufficient deviations of supply voltages from nominal, for example, to accommodate variations in resistive drops in wiring and fuses, as well as tolerances in supplies. Typical specified tolerances are ±5% or ±10% for subsystems for commercial applications.

In real applications one can expect to find noise on the power leads caused by variations in the loads on the power supplies and by switching noise generated within the supplies. The allowed levels and allowed frequency content must be specified. Typical values are of the order of 10–100 mV with components at frequencies of up to 100 kHz depending upon the types of filtering and wiring configurations used.

Incorrect voltage and overvoltage protection is another consideration when designing a subsystem. A good design will not sustain permanent damage if one or more of the power supplies is turned off for an extended period of time, for example, if there is a supply failure. The need for the ability to withstand abuse such as power supplies hooked up to the wrong leads or surge voltages far in excess of nominal is a matter of judgment depending upon the application and the value vs. cost of such protection.

2.3.2. Environment

Various applications of fiber systems place different environmental constraints on the design.

For some intrabuilding applications, with air conditioning and adequate power available, one can often assume that the ambient around the equipment will be limited to say 30°C maximum temperature, under normal conditions. If the equipment being interconnected by the fiber link is itself temperature sensitive (e.g., some types of computers) one could design and specify the link to operate properly only within a narrow temperature range (assuming that such a sacrifice in performance is justified by some cost or reliability benefit). At the other extreme, outside plant equipment on telephone poles and in manholes must operate satisfactorily with wide swings in temperature and humidity. The cost and maintenance of equipment and power limitations make strict environmental control in these applications impractical.

As mentioned in Section 2.3.1, dc power busses and signal lines can be subjected to overvoltages due to lightning and power crosses in some applications. When local power is provided from a remote location by copper pairs paralleling the fiber (e.g., in an outside plant application), the design of the equipment must take these potential overvoltages into account. Overvoltage spikes can even occur on the signal leads and grounds between frames in a building. One of the advantages of fiber systems is the ability bridge such potential differences between frames. However, the system design must not overlook problems such overvoltages can cause.

2.3.3. Operational Requirements

Inaccessibility of equipment or system availability objectives place requirements on the reliability of components. Some applications allow for a reasonable amount of "down time" and provide sophisticated technicians to maintain the system. Other applications require very high availability and simple maintenance.

In some applications one can allow for numerous adjustments to be made to make the link operational. In other applications, the components must essentially snap together, assuming little understanding of their operation on the part of the installer. For example, in many applications the fiber transmitter and receiver must operate with a wide range of possible optical attenuations between them. This places a constraint on the dynamic range of the receiver, which results in trade offs in the design. This also puts a requirement on the variation in transmitter output with temperature and time since such variations could use up unnecessarily large amounts of receiver dynamic range.

Subsystems (1-6)

3.1. Overview

In this chapter we shall discuss the subsystems which make up typical fiber optic transmission links. In particular, we shall discuss transmitters, receivers, and fibers (in terms of their input/output properties). In order to treat the pieces separately, we shall make use of the artifices of an ideal transmitter, receiver, or fiber, to be defined below. Thus, for example, we can examine the details of an actual transmitter by imagining it as part of a system incorporating an ideal fiber and receiver.

3.2. Ideal Subsystems

The ideal transmitter subsystem would convert an incoming electrical waveform into an optical output power which exactly tracks the incoming signal. Thus, if the incoming waveform is $m(t)$ (volts) the outgoing optical power $p(t)$ (watts) would be given by

$$p(t) = P_0 [\beta + \gamma m(t)] \qquad \text{(watts)} \qquad (3.2.1)$$

where β and γ are chosen so that $p(t)$ is, of course, always positive.

The ideal fiber would accept as its input the output power from the transmitter. The fiber output power would be an exact replica of the fiber input power except for an attenuation factor.

The ideal receiver would convert incident optical power to an output voltage waveform which is an exact replica of the input power waveform except perhaps for an ac coupling filter which might be specified at the receiver output.

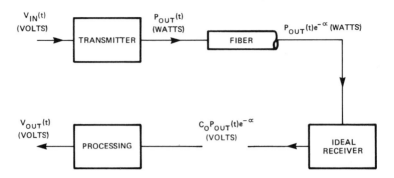

Figure 3.1. Transmitter subsystem plus ideal fiber plus receiver subsystems.

3.3. Transmitter Subsystems

Figure 3.1 shows a transmitter subsystem as part of a system incorporating an ideal fiber and receiver. The transmitter input waveform is the voltage $v_{in}(t)$. We can begin our study of transmitter subsystem design by considering input/output characteristics.

Figure 3.2 shows a binary clocked input waveform and an ideal output waveform. Also shown are a number of "actual" output waveforms showing various types of degradations (some exaggerated). Waveform 3 shows pulse pattern-dependent amplitude variations and pulse width variations. Waveform 4 shows intersymbol interference caused by an inadequate rise time and fall time. Waveform 5 shows output pulse overshoot and ringing. Waveform 6 shows an inadequate extinction ratio. Waveform 7 shows noise on the pulses. In any practical transmitter these defects are all present to some extent. A typical transmitter specification might look like the following:

Transmitter Specification (Specimen)

1. All specifications will be valid over the temperature range -20 to $60°$C (except lifetime).

2. The transmitter subsystem will operate with applied voltages of $+5$ and -5.2 V $\pm 5\%$, with a maximum current drain of 250 mA on either supply. The signal input interface will be balanced ECL (Emitter-Coupled Logic).

3. The transmitter will incorporate a light-emitting diode or a laser.

4. If a light-emitting diode is used, the center wavelength of emission will be 830 ± 10 nm and the spectral width (FWHM) at $20°$C will be less than 50 nm. If a laser is used the center wavelength will be 830 ± 10 nm and the spectral width will be less than 1 nm.

5. The transmitter output will be from a graded index optical fiber having a numerical aperture (N.A.) of 0.2 ± 0.02, a core diameter of 62.5 ± 3 μm, and a cladding diameter of

125 ± 6 μm. The core will be centered within the cladding with an axial offset of no more than 3 μm.

6. The transmitter output in the "on" state will be between −18 and −15 dBm for an LED source at the beginning of life. End of life for an LED source will be defined as an output below −21 dBm. The output in the "on" state will be −1.5 dBm ± 1.5 dB for a laser source. End of life for a laser source will be defined as an increase in subsystem current drain beyond the limits stated in 2, for feedback-controlled devices, or a decrease in output to below −6.0 dBm in the "on" state, whichever occurs first.

7. Pattern-dependent effects will be tested by driving the transmitter with the output of a $2^{17} − 1$ pseudorandom sequence generator operating at a 20-MHz clock rate, NRZ. Pulse amplitude will not vary with the pattern by more than ±5%. Rise and fall times will be less than 10 ns (10%–90%). Pulse width will not vary with the pattern by more than ±5% (FWHM). Overshoot will be less than 10%. Oscillations will not occur at frequencies below 200 MHz (laser sources). The extinction ratio will exceed 10:1.

8. The mean lifetime of the source at 20°C ambient will exceed 100,000 hr. The mean lifetime at 60°C ambient will exceed 1000 hr.

The above "typical" specification was presented to give the reader a feeling for the parameters of interest and what typical requirements might be. An actual specification might give considerably more detail about how some of the parameters are to be measured, and certainly could specify different requirements on the tolerances.

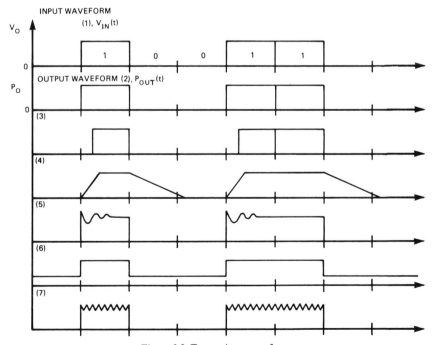

Figure 3.2. Transmitter waveforms.

3.3.1. Spectral Considerations

In many applications one can consider the total output power of the transmitter source as a single parameter; however, sometimes one must take into account the spectral content of the source. For example, a typical light-emitting diode emits light over a band of wavelengths of approximately 5% of the center wavelength in width. These different wavelengths travel at different group velocities in the fiber (material dispersion). Suppose the "material dispersion" in the fiber is 100 ps of delay difference per nanometer of spectral width per kilometer of transmission. This means that for an LED with a spectral width of 50 nm, light at one end of the spectrum will be delayed 5 ns/km relative to light at the other end of the spectrum. More precisely, suppose the transmitter emits light with a Gaussian-shaped spectrum of the following form:

$$S(\lambda) = \left[\frac{1}{(2\pi)^{1/2} \Lambda} \right] e^{-(\lambda - \lambda_0)^2/2\Lambda^2} \tag{3.3.1}$$

where λ_0 is the spectral center wavelength (nanometers) and Λ is the rms spectral width (nanometers).

Suppose now that the source is modulated with an input current that is a Gaussian-shaped pulse, producing a Gaussian-shaped output power pulse of the following form:

$$p(t) = \frac{1}{(2\pi)^{1/2}\sigma} e^{-t^2/2\sigma^2} \tag{3.3.2}$$

where σ is the rms source output pulse width (nanoseconds) and where $p(t)$ is the total source output power vs. time, integrated over all wavelengths which are emitted.

Suppose now that this power propagates through a fiber that has a wavelength-dependent propagation delay, $d(\lambda)$. The output pulse from the fiber, neglecting attenuation, can be obtained by integrating the output as a function of wavelength as follows:

$$p_{\text{fiber output}}(t) = \int S(\lambda) \frac{1}{(2\pi)^{1/2} \sigma} e^{-[t - d(\lambda)]^2/2\sigma^2} d(\lambda) \tag{3.3.3}$$

If the wavelength-dependent delay is approximately a linear function of wavelength, i.e.,

$$d(\lambda) = d(\lambda_0) + D(\lambda - \lambda_0) \tag{3.3.4}$$

where $D = \delta L$ and L is the fiber length, then the fiber output pulse can be obtained by substituting (3.3.1) and (3.3.4) into (3.3.3):

$$p_{\text{fiber out}}(t) = \frac{1}{(2\pi)^{1/2}\,\gamma}\, e^{-[t - d(\lambda_0)]^2/2\gamma^2} \qquad (3.3.5)$$

where $\gamma^2 = \sigma^2 + \Lambda^2 D^2 = \sigma^2 + \Lambda^2 \delta^2 L^2$.

We see that the dispersion in the fiber, δ (ns nm^{-1} km^{-1}), results in an increase of the rms width of the pulse to a value which is the square root of the sum of the square of the input pulse width, σ, and the square of a term proportional to the dispersion δ, the source bandwidth, Λ, and the fiber length, L.

At 0.8 μm wavelength δ is approximately 100 ps nm^{-1} km^{-1} in typical fibers made of silica. The spectral width of a typical GaAs LED is about 17 nm (rms) at room temperature. There is a decrease of δ with increasing wavelength, reaching a minimum (zero) at about 1.3–1.4 μm. This material dispersion null and the lower loss of fibers at this wavelength motivate research in sources and detectors for the 1.3–1.5-μm region.

All of the above assumes that the transmitter output waveform is independent of wavelength over the band of wavelengths it emits.

It has been observed that under some conditions light-emitting diodes can exhibit a phenomenon called "chromatic delay" or "chirping." As the device is modulated, there is a delay between the modulating waveform and the output power waveform that varies with wavelength within the emitted spectrum of the device. Thus, the output power as a function of time and wavelength for an input modulation $m(t)$ takes the form

$$p_{\text{out}}(t, \lambda) = p_0 \{ m[t - d_c(\lambda)] \} \qquad (3.3.6)$$

where $d_c(\lambda)$ is the wavelength-dependent (chromatic) delay.

When light from such a source propagates through a fiber with material dispersion, the wavelength-dependent delay of the fiber adds to the wavelength-dependent chromatic delay of the source. Since it is possible for the delay vs. wavelength of the source to be opposite to the incremental delay vs. wavelength of the fiber, fiber material dispersion can actually result in a narrowing of the pulse in transmission (dechirping). For most practical applications, the chromatic delay effect is generally neglected.

3.3.2. Circuits for LED Transmitters†

A light-emitting diode is a *pn* junction with a specified input capacitance and resistance vs. applied bias current. Figure 3.3 shows some typical diode characteristics. We see that one wishes to apply a current of about

† See Reference 6 in Chapter 5.

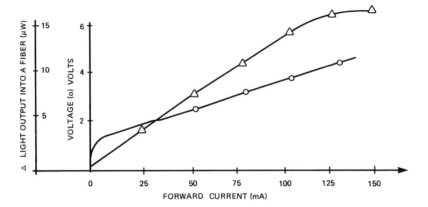

Figure 3.3. Typical light-emitting diode characteristics.

100 mA to turn the diode on. This will produce a voltage drop across the diode of about 2–5 V depending upon the diode series resistance. Figures 3.4 and 3.5 show two typical driver circuits which can interface a standard TTL (Transistor–Transistor Logic) and a standard balanced ECL source to an LED. Assuming that the circuit has sufficient current and voltage capability to drive the device at slow speeds, the most significant parameter for a given driver circuit is usually the speed of the circuit–LED combination.

There are some fundamental limitations on the speed of response of an LED device. The LED emits light when holes and electrons combine in the junction. There is a parameter associated with an LED, with a particular doping level profile and made from specific materials, called the spontaneous recombination lifetime. This parameter is the average time it takes for an electron injected into the junction to find a hole and to combine with the hole to emit light or heat. Suppose the current being injected into the device

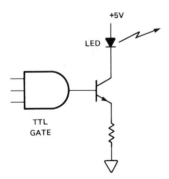

Figure 3.4. Simple TTL interface LED driver.

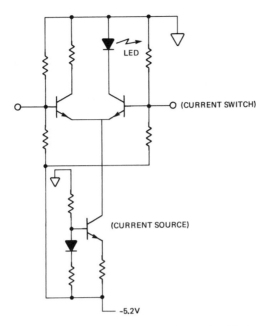

Figure 3.5. ECL interface balanced driver (simplified).

is represented by the parameter $i(t)$. The number of carriers in the junction volume is represented by the following equations:

$$\frac{\partial n(t)}{\partial t} = \frac{i(t)}{e} - \frac{n(t)}{\tau_{sp}} \tag{3.3.7}$$

where τ_{sp} is the spontaneous recombination lifetime and $n(t)$ is the number of excess carriers in the junction volume, with e the electron charge.

The rate at which light is emitted is proportional to the excess carrier recombination rate and is given by

$$p(t) = \eta hf\, n(t)/\tau_{sp} \tag{3.3.8}$$

where hf is the energy in a photon and η is an overall efficiency factor which takes into account the fraction of recombining carriers which emit light rather than heat, and also the fraction of emitted light which is capable of reaching the emitting surface before being reabsorbed. The relationship between the emitted light and the current injected into the junction is therefore given by

$$p(t) = \frac{\eta hf}{e} \int i(t')\, e^{-(t-t')/\tau_{sp}}\, dt' \tag{3.3.9}$$

Thus the LED has an output power waveform which is a filtered version of the input current $i(t)$, with bandwidth parameter $B = 1/(2\pi\tau_{sp})$.

Of course in (3.3.9) $i(t)$ represents the current injected into the junction. In order to achieve high modulation rates, one must use a drive circuit which has a sufficiently low impedance to overcome the LED capacitance, (often more than 100 pF at zero bias), so that the transmitter speed is LED carrier-lifetime-limited rather than circuit limited.

It should be pointed out that one can extend the bandwidth capabilities of an LED by using driver circuits which enhance high frequencies (equalize). However, such drivers tend to require large amounts of power and often must be matched (adjusted) to a given LED. Thus, it is preferable to use LED devices with sufficiently low capacitance and sufficiently low recombination lifetimes (if available) rather than complex equalization techniques in the driver.

When designing LED drivers for high-frequency operation, one must take into account the variations in the capacitance and forward resistance of the device with forward current. The resulting nonlinearities can lead to problems in some driver designs including pattern-dependent pulse shapes and intersymbol interference. This problem can be reduced using the circuit shown in Figure 3.6 which provides a dc bias for the LED, at a sacrifice of extinction ratio.

The large currents required to modulate LEDs can cause problems associated with powering transmitter subsystems. The power supply leads must be very well filtered to avoid feedback from the transmitter into other system components. One way to reduce this effect is to use a balanced driver as shown in Figure 3.5. This sacrifices power consumption for increased modulation speed capabilities and reduced power line noise.

Since the LED light output vs. current input characteristic is fairly linear up to the point of saturation, multilevel drivers for LEDs and analog drivers for LEDs are not particularly complex. However, if very high

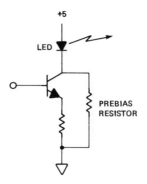

Figure 3.6. Prebiased LED driver.

linearity or very precise level spacing is required, then various feedback, feedforward, or predistortion techniques must be employed.

Figure 3.7 shows an LED driver with feedback control. As shown, a portion of the emitted light is captured by a local light detector, amplified, and compared to the drive signal. The equations which govern the linearity improvement provided by feedback can be derived as follows:

Let the power and drive current relationship of the LED be approximated as follows:

$$p(t) = \alpha_1 i(t) + \alpha_2 i^2(t) = \alpha_1 i(t) \left[1 + \frac{\alpha_2}{\alpha_1^2} \alpha_1 i(t) \right] \qquad (3.3.10)$$

where

$$\frac{\alpha_2}{\alpha_1^2} \left[\alpha_1 i(t) \right] \ll 1 \quad \text{for } \alpha_1 i(t) \text{ values being used}$$

If the fraction of light output power captured by the local detector is γ, if the detector–amplifier responsivity transimpedance is A_1 V W^{-1}, and if the drive-amplifier gain is A_2 A V^{-1}, then we have

$$i(t) = A_2 \left[m(t) - A_1 \gamma p(t) \right] \qquad (3.3.11)$$

$$i(t) = \frac{-\alpha_1 + \left[\alpha_1^2 + 4\alpha_2 p(t) \right]^{1/2}}{2\alpha_2}$$

$$= \frac{\alpha_1}{2\alpha_2} \left\{ -1 + \left[1 + \frac{4\alpha_2}{\alpha_1^2} p(t) \right]^{1/2} \right\} \qquad (3.3.12)$$

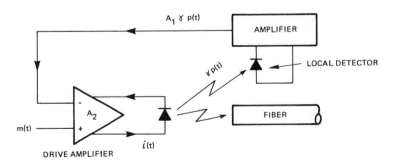

Figure 3.7. Feedback-controlled LED driver.

Taking terms up to order $\left[(\alpha_2/\alpha_1^2)\,p(t)\right]^2$ we get

$$i(t) \approx \frac{\alpha_1}{2\alpha_2}\left\{-1 + 1 + \frac{2\alpha_2\,p(t)}{\alpha_1^2} - 2\left[\frac{\alpha_2\,p(t)}{\alpha_1^2}\right]^2\right\} \qquad (3.3.13)$$

Then substituting (3.3.13) into (3.3.11) we obtain

$$\frac{p(t)}{\alpha_1} - \frac{p^2(t)}{\alpha_1}\frac{\alpha_2}{\alpha_1^2} = A_2\left[m(t) - A_1\gamma p(t)\right] \qquad (3.3.14)$$

$$p^2(t)\frac{\alpha_2}{\alpha_1^3} - p(t)\left[A_2A_1\gamma + \frac{1}{\alpha_1}\right] + A_2 m(t) = 0$$

$$p(t) = \left\{A_2A_1\gamma + \frac{1}{\alpha_1} - \left[\left(A_2A_1\gamma + \frac{1}{\alpha_1}\right)^2 - \frac{4A_2\alpha_2 m(t)}{\alpha_1^3}\right]^{1/2}\right\}\left(\frac{2\alpha_2}{\alpha_1^3}\right)^{-1}$$

$$p(t) = \left[\frac{A_2 m(t)}{A_2A_1\gamma + 1/\alpha_1}\right]\left[\frac{A_2\alpha_2 m(t)}{\alpha_1^3(A_2A_1\gamma + 1/\alpha_1)^2} + 1\right]$$

Now if we fix the level of the linear part of the modulation, setting $A_2/(A_2A_1\gamma + 1/\alpha_1) = C$ (a constant), we obtain

$$p(t) = C\,m(t)\left[1 + \frac{(\alpha_2/\alpha_1^2)\,C\,m(t)}{1 + A_2A_1\alpha_1\gamma}\right] \qquad (3.3.15)$$

Thus since C, α_2, and α_1 are fixed for a given linear modulation level and a given LED, we obtain a nonlinear term proportional to $(1 + A_2A_1\alpha_1\gamma)^{-1}$, where $A_2A_1\alpha_1\gamma$ is the gain around the feedback loop. A loop gain of 10 reduces the nonlinear term by a factor of 11 compared to no feedback.

All of the above assumes that the power launched into the fiber in Figure 3.7 is closely correlated with the power captured by the local detector. The validity of this assumption depends upon the nature of the beam splitter.

The light-emitting diode emits energy in a number of spatial modes, corresponding to the large emission solid angle relative to the solid angle occupied by a single diffraction-limited plane wave emitted by an area equal to the LED emission area (more on this in Section 4.1.1). The light power vs. drive current characteristics of different modes are not necessarily the same. Thus an optical tap or beam splitter that diverts energy from selected modes to the local detector may result in a linearization of the light vs. current characteristics of only those modes, and not the overall diode output. For linearizing an LED with local feedback, it is important to use a beam splitter or tap that samples the same mode distribution as the fiber captures. The more linearization one tries to obtain, the more difficult this is to achieve.

Another problem with feedback linearization is the delay around the feedback loop, which limits the bandwidth available in a feedback-controlled analog transmitter.

Other approaches to linearization are to predistort the drive current in an attempt to cancel the LED nonlinearities; to use a small modulation index with a selected bias point; and to use various forms of feedforward techniques. All of these approaches assume a certain stability of LED characteristics with time and temperature—which may or may not be valid.

3.3.3. Circuits for Laser Transmitters

One significant difference between a laser and an LED is the threshold behavior of the light out vs. drive current as shown in Figure 3.8. As the drive current is increased from zero the laser first behaves as a light-emitting diode with very little power coupled into the fiber, because the light emitted is spatially incoherent. (Coupling light sources to fibers will be discussed further in Section 4.1.1.) At sufficiently high current densities, a population inversion is produced within the laser cavity resulting in a domination of stimulated emission over absorption. Thus the amplitudes of the lowest loss modes of the laser cavity begin to grow until these fields saturate. Above the threshold current, i_{th}, the laser can emit large amounts of power into a

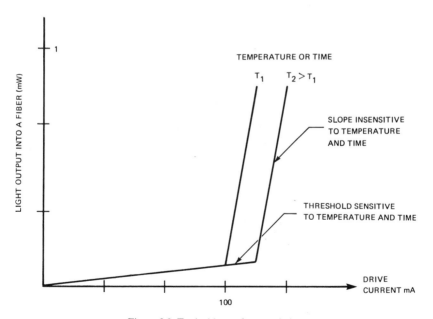

Figure 3.8. Typical laser characteristics.

fiber, because most of its total output is contained in just a few spatial modes.

To use a laser, it is often desirable to "prebias" the device near the threshold current i_{th}, in order to avoid the time delays necessary to build up the carrier densities, $n(t)$, within the device to the levels associated with the threshold current. That is, if n_{th} is the carrier density within the cavity at threshold and τ_{sp} is the spontaneous recombination lifetime, then for an applied current i_{drive} turning on at time $t = 0$, the delay until lasing occurs is derived as follows:

$$\frac{\partial n(t)}{\partial t} = \frac{i_{drive}}{e} - \frac{n(t)}{\tau_{sp}}, \qquad t \geq 0 \qquad (3.3.16)$$

$$n(t) = \frac{\tau_{sp}\, i_{drive}}{e}(1 - e^{-t/\tau_{sp}})$$

therefore since

$$i_{th} = n_{th} e / \tau_{sp}$$

we have

$$t\,|\,[n(t) = n_{th}] = \tau_{sp} \ln\left[1 - (i_{drive}/i_{th})\right]$$

To minimize this turn-on delay, one can drive the laser (from zero current) with a signal current significantly in excess of the threshold i_{th}. However, once the laser carrier density exceeds threshold density, such large currents, if continuously applied, are likely to burn out the device by producing excess field intensities at the facets, or accelerating the growth of crystal defects. Thus, as an alternative, by biasing the laser slightly above threshold, one can eliminate this turn-on time delay. An additional benefit of prebiasing the laser is that the incremental current needed to drive the laser from threshold to its maximum safe output can be much smaller than the bias (say, 20–30 mA compared to 100–200 mA). Thus the high-speed modulating signal for a laser with prebias can be much smaller than the signal needed to modulate an LED. This is important when fast drivers are needed. A disadvantage to prebias is the increased power consumption compared to a circuit which drives the laser on from zero current.

Some moderate-speed driver circuits take advantage of the capabilities of some lasers to produce large, low-duty-cycle pulses of optical power, without damage, by driving the laser with large, low-duty-cycle current pulses. The net efficiency of conversion of electrical power to optical output can be higher with such circuits. Moreover, such circuits may not have to be carefully adjusted and/or temperature compensated, to account for the variation in the laser threshold with temperature, provided laser damage is not caused by the range of large, low-duty-cycle output pulses which may occur as the threshold varies.

The variation of the threshold with temperature and time is a significant problem with any circuit which attempts to bias the laser at a fixed level at or relative to the threshold. Various means have been used to stabilize the prebias current, with varying amounts of success.

One approach is to monitor the laser output with an optical tap and a local detector as shown in Figure 3.9. A feedback circuit is used to control the prebias to cause the average output power of the laser to be constant. A problem with this approach is that it assumes that the average value of the drive signal voltage waveform is a constant. If the drive signal voltage is temporarily removed, the feedback control circuit will cause the prebias to increase so that the laser is turned on with a constant average output value equal to the average output value it had with the signal drive voltage applied. When the signal drive voltage is reconnected, the temporary high setting of the prebias may cause the laser to burn out when the signal drive voltage assumes its high state.

A better approach is shown in Figure 3.10.[7] Here the average output power of the laser is compared to the average value of the signal drive voltage. If the signal drive voltage is removed, its average value will go to zero, forcing the average laser output to zero, avoiding the problem described above. This circuit assumes that the variation of the laser light-current characteristic with temperature and time is as shown in Figure 3.8. Thus as temperature or time varies, one can compensate with changes in prebias rather than with the incremental drive current. Evidence to date indicates that the slope of the curve (differential quantum efficiency) is

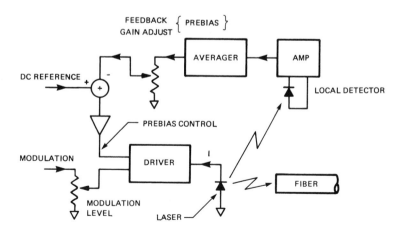

Figure 3.9. Simple laser bias stabilization circuit.

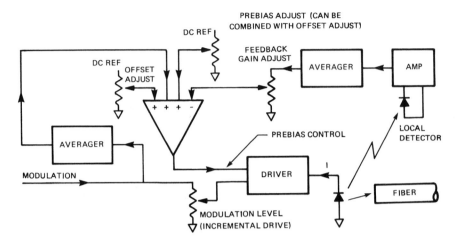

Figure 3.10. Improved prebias stabilization circuit.

indeed approximately constant, for a given laser, with temperature and time as shown in Figure 3.8.

Circuits like the one shown in Figure 3.10 require not only a large number of components, but also careful alignment. One has to set the incremental drive current, the prebias, and at least one gain and one offset (to compensate for the fact that the input signal drive voltage waveform and the optically derived feedback voltage waveform have a relative offset as well as a scale difference). These settings are interactive, resulting in a complex alignment procedure.

There are some special applications for optical pulses which are very narrow in width (sometimes less than 100 ps) and having a low repetition rate (a few tens of kilohertz). Such pulse streams can be generated by taking advantage of the fact that some powerful injection lasers tend to emit narrow pulses of light even under dc drive current conditions. A transmitter for such a narrow pulse emitter is shown in Figure 3.11. The capacitor C_1 charges through R_1 to store a fixed charge Q. When C_1 is charged, the transistor collector–emitter voltage exceeds V_{CEV}, but is less than V_{CER} (collector–emitter reverse breakdown voltages for low and moderate impedance between base and emitter respectively). Thus when a voltage trigger pulse is applied to the transistor base, the transistor will rapidly turn on, dumping the charge Q through the emitter resistor and the laser diode. This pulser takes advantage of the turn-on delay of the laser and the fact that the laser will produce narrow spikes of light output once its threshold is exceeded. The voltage V_1 is adjusted so that Q is just large

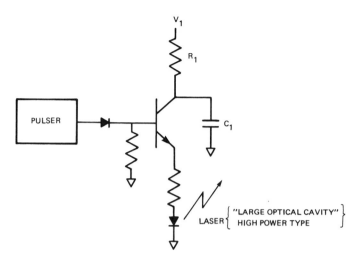

Figure 3.11. Fast pulser.

enough to charge the laser to threshold, and to produce one narrow spike. The width of this spike is typically less than 100 ps even though the current pulse may last several nanoseconds.

3.4. Receiver Subsystems [1, 3]

The function of a receiver in a fiber optic link is to convert light power to an electrical signal. The receiver consists of a detector, a preamplifier and a main amplifier as shown in Figure 3.12. An ideal receiver would produce a waveform $V_{out}(t)$ at its output which is an exact replica of the input power waveform $p(t)$. Actual receivers will introduce the following effects: noise, linear filtering, and distortion. In this section we shall char-

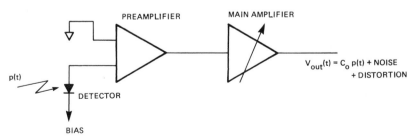

Figure 3.12. Typical receiver.

acterize the various sources of noise and their effects on the receiver performance. We shall also describe the limitations on bandwidth and distortion in practical receivers.

3.4.1. Fundamental Sources of Noise

In classical high-frequency and microwave circuits, the most fundamental source of noise generally considered is thermal background noise which enters the receiver along with the signal through the antenna. The level of this noise is generally given by kT watts per hertz, where k is Boltzmann's constant and T is the noise temperature of the antenna. When optical frequencies are being used one must make use of a more exact expression for background noise given by

$$N_B = \frac{hf}{e^{(hf/kT)} - 1} \qquad \text{(watts/hertz)} \qquad (3.4.1)$$

where hf is the energy in a photon.

At microwave frequencies and below, $hf \ll kT$, and the expression (3.4.1) reduces to kT. However, at 1 micrometer wavelength hf is about 2×10^{-19} and kT (at room temperature) is about 4×10^{-21}. Thus at optical wavelengths the background radiation given by (3.4.1) is very small, even if very large optical bandwidths are allowed to enter the receiver. As a consequence, another effect (usually neglected at microwave frequencies compared to background noise) becomes the fundamental noise limitation. This effect is usually referred to as quantum noise. Quantum noise manifests itself in several possible ways depending upon the type of optical detection process being used. For photodiodes it manifests itself in the statistics of photodetection.

When optical power is incident upon a back-biased photodiode hole–electron pairs are generated within the device. These hole–electron pairs separate under the influence of the fields within the photodiode to produce a displacement current. On the average, the number of hole–electron pairs created per second, $\langle n(t) \rangle$, is proportional to the incident light power

$$\langle n(t) \rangle = p(t)\eta/hf \qquad (3.4.2)$$

where $p(t)$ is the light power, in watts, η is the detector quantum efficiency, and hf is the energy in a photon.

If we define $\eta p(t)$ as the "detected power" (power actually converted to current), we see that the number of hole–electron pairs generated per second on the average is equal to the average number of detected photons incident per second.

Equation (3.4.2) represents the average rate at which hole–electron pairs are generated. In any interval of time T, if the detected power incident upon the detector is $p_d(t)$, the exact number of electron–hole pairs which will be generated is not precisely predictable. The number generated is statistically predicted by the following probability distribution:

$$p_N(n) = \Lambda^n e^{-\Lambda}/n! \qquad (3.4.3)$$

where

$$\Lambda = (1/hf) \int p_d(t)\,dt \triangleq E_d/hf$$

and where N is the number of hole electron pairs generated, $p_N(n)$ is the probability of $N = n$, E_d is the energy detected in interval T, and $\triangleq$ means "by definition, equal to".

The fact that one cannot exactly predict how many electron–hole pairs the power incident upon the detector will generate, even though the power is known, represents a type of noise referred to as quantum noise.

An example of how quantum noise limits the performance of an optical receiver can be derived as follows.

Consider a communication system where binary digits are transmitted at a rate B, bits/second, by turning an optical transmitter on or off. The energy in a transmitted pulse is either 0 or E_T (joules). Thus the transmitter extinction ratio is infinity. At the receiver, attenuated pulses are received having energy 0 or E_d (joules). The receiver detector produces no dark current, so no hole–electron pairs can be generated in the absence of detected light. The receiver preamplifier contributes such a low additive noise that even a single hole–electron pair generated in the receiver detector will produce an observable response. Therefore, if one or more hole–electron pairs are generated, the receiver output is interpreted as a "mark" (optical pulse present). If no hole–electron pairs are generated in a pulse interval, the receiver output is interpreted as a "space" (no optical pulse).

Since there is no dark current, no hole–electron pairs can be generated if no optical pulse is present. The only way an "error" can be made is if an optical pulse E_d is indeed present at the receiver, yet no hole–electron pairs are generated. The probability of such an occurrence is given by (3.4.3) with $n = 0$. That is,

$$P_E = p_N(0) = e^{-E_d/hf}$$

If we wish to have a probability of error of one in a billion, we must set $P_E = 10^{-9}$, which corresponds to $E_d = 21hf$.

In other words, if the average number of photons contained in the optical pulse is 21, only one time in a billion will no hole–electron pairs be generated and an error made. The average power detected is given by the average energy detected per pulse multiplied by the pulse rate B and divided by 2 (for half "marks" and half "spaces"). Thus the minimum detectable power level for a 10^{-9} error rate is given by

$$p_{min} = 10.5hfB \qquad (3.4.5)$$

For $hf = 2 \times 10^{-19}$ (i.e., at about 1 μm wavelength) and $B = 10$ Mbaud/s we obtain $p_{min} = 2 \times 10^{-11}$ W or -77 dBm. This is called the quantum limit for digital detection at 10 Mbaud/s with a 10^{-9} error rate because all aspects of the system are ideal, and the performance is limited only by the statistics of photodetection. We shall see later that practical receivers can be built with sensitivities which approach 13 dB of the quantum limit.

3.4.2. Noise in Amplifiers

As described above, the fundamental noise limitation in optical fiber systems is quantum noise associated with the statistics of the detection process. In order to achieve quantum limited performance, individual hole–electron pairs generated in the detector must be observable in the background noise added by the preamplifier. It is easy to show that the displacement currents of individual electron–hole pairs are not in fact observable, and that only the cumulative effect of the superposition of many displacement currents from many pairs is observable. Consider, for example, the simple pin detector–amplifier combination shown in Figure 3.13. The input to the amplifier is a 50-Ω resistance to ground, and for the sake of this argument, the amplifier is assumed to have a noise at its output dominated by the Johnson noise of the 50-Ω input resistance. The bandwidth of the amplifier is assumed to be 10 MHz and the gain of the amplifier is 100 (actually irrelevant to this discussion). The equivalent circuit of the amplifier–detector combination is shown in Figure 3.14. The capacitance C_d is the diode junction capacitance, typically a few picofarads. The

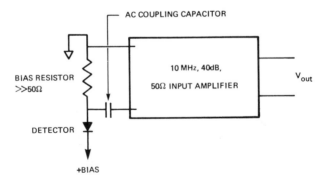

Figure 3.13. Simple detector–amplifier.

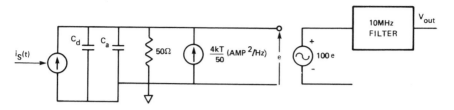

Figure 3.14. Equivalent circuit of simple detector–amplifier.

detector photodetection mechanism is represented by a current source which produces pulses of area e coulombs each time a hole–electron pair is generated inside the device. These current pulses are assumed narrow in time compared to the reciprocal bandwidth of the amplifier. Thus they are effectively impulses (Dirac delta functions) of area 1.6×10^{-19} C. The amplifier has an input capacitance C_a and a shunt resistance of 50 Ω. In parallel with this resistance is a noise source of spectral density $4kT/50$ A^2 Hz^{-1}, where $kT = 4 \times 10^{-21}$. This noise source represents the Johnson noise of the physical input resistance.

Since the bandwidth of the amplifier is 10 MHz, we can neglect the admittance of the capacitors at the input compared to the admittance of the 50-Ω resistor. To answer the question as to whether individual electron–hole pairs are observable, we must calculate two quantities. One is the rms noise level at the receiver output. The other is the peak of the output response produced when an impulse of area e coulombs is emitted by the detector equivalent current source. The rms thermal noise at the receiver output is given by

$$n_{\text{out rms}} = \left[\frac{4kT}{50} \times (50)^2 \times 100^2 \times 10^7 \right]^{1/2} \cong 2.8 \times 10^{-4} \, \text{V} \quad (3.4.6)$$

The response at the output to an impulse of current of area e emitted by the detector has area $e \times 50$ (ohms) $\times 100$ (amplifier gain). Its peak height is roughly given by its area divided by its duration in time. Since the amplifier has bandwidth B, the pulse has duration roughly $1/B$. Thus the peak response to a single hole–electron pair at the output is given by

$$v_{e \text{ peak}} = 1.6 \times 10^{-19} \times 50 \times 100 \times 10^7 = 8 \times 10^{-9} \, \text{V} \quad (3.4.7)$$

We see from (3.4.7) and (3.4.6) that for a 50-Ω 10-MHz amplifier, the rms noise at the amplifier output is about 35,000 times as big as the response produced by a single hole–electron pair. Thus individual hole–electron pairs are not observable. We can define a quantity Z as the ratio of rms output noise to the response to a single hole–electron pair. In this case Z

is approximately 35,000. In general Z is given by (for an amplifier limited by the noise of a physical resistor R at its input):

$$Z_{\text{resistor}} = [4kTB/R]^{1/2}/eB \qquad (3.4.7a)$$

This parameter Z is a figure of merit for optical fiber system preamplifiers. It is more meaningful than several conventional amplifier measures such as noise figure, and receiver sensitivity measures such as noise equivalent power. In the case of noise figure, since the detector is a current source in parallel with a capacitor, there is no source resistance and therefore no source noise to measure the amplifier noise against. That is, except for dark current which varies markedly from one detector to another, there is no effective source noise resistance to normalize the amplifier noise to. In effect, any amplifier would have an enormous noise figure compared to the detector thermal noise. Noise equivalent power can be a useful measure of receiver performance provided it is properly defined. Many receiver specifications still normalize noise equivalent power to "root bandwidth," which we shall see is inappropriate in most fiber applications. The parameter Z can be unambiguously defined as a function of the detector capacitance and the bandwidth being used. For most receiver sensitivity calculations it is the most fundamental thermal noise parameter, as will be seen below. For a simple 50-Ω amplifier, Z is about 35,000 at 10 MHz and varies with the receiver bandwidth as (bandwidth)$^{-1/2}$ over the range of frequencies where the thermal noise of the physical 50-Ω resistor dominates all other amplifier noises. We shall see that using preamplifiers specifically designed to work with photodetectors, values of Z of less than 1000 can be obtained.

Before describing ways to obtain lower values of the parameter Z we should explain how Z can be measured in actual amplifier detector packages, and we should describe the effect of Z on receiver performance.

Figure 3.15 shows an experimental setup for determining Z. The detector is illuminated with an optical pulse train whose individual pulses are

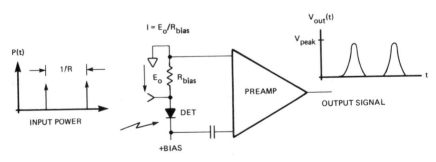

Figure 3.15. Measuring Z.

much narrower than the reciprocal of the receiver bandwidth. The spacing between pulses is much larger than the reciprocal of the receiver bandwidth. The light power is removed temporarily, and two quantities are measured: the rms noise level at the amplifier output and the average current flowing through the detector. The light power is now applied and increased in level until pulses are clearly observable at the receiver output. Each of these output pulses represents the cumulative response of many thousands of hole–electron pairs produced in the detector in response to an optical pulse. The amplitude of a receiver output pulse is measured using an oscilloscope. The number of hole–electron pairs contributing to an output pulse is by definition $Z \times$ (peak height of an output pulse/rms noise). The number of hole–electron pairs in an output pulse is also given by the increase in the average detector current (when light is applied) divided by (pulse repetition rate × the charge in an electron). Thus algebraically

$$Z = \left(\frac{\Delta I}{Re} \right) \left(\frac{v_{\text{peak}}}{v_{n \text{ rms}}} \right)^{-1} \tag{3.4.8}$$

where ΔI is the increase in average detector current when light is applied, R is the pulse repetition rate, e is the electron charge, v_{peak} is the peak observed amplifier output pulse, and $v_{n \text{ rms}}$ is the amplifier rms output noise.

In the above measurement of Z it does not matter whether an APD (avalanche photodiode) or pin detector is used; what the quantum efficiency is; what the dark current is (provided dark current noise is negligible compared to amplifier thermal noise); or what the wavelength is. What is important is that the detector used to measure Z has a junction and interconnection capacitance which is specified, or comparable to the capacitance anticipated in actual operation. Also important is that the on–off ratio of the light pulse stream be high enough, or known accurately enough, so that the contribution of dc (unmodulated) light to ΔI can be accounted for.

The relevance of parameter Z to digital systems is easy to calculate for large Z values and when PIN detectors are being used. For example, when Z is large and without avalanche gain one can neglect the Poisson statistics of photodetection in calculating the optical power level required to achieve a fixed error rate in a binary optical link. From standard communication theory,[8] and assuming that the amplifier thermal noise has Gaussian statistics, it follows that to achieve a 10^{-9} error rate the receiver output response to an incident optical pulse must be 12 times as large as the rms value of the noise at the receiver output. Since the rms noise is Z times as large as the receiver output response produced by a single electron emitted by the detector, it follows that the required number of hole–electron pairs which must be produced by the incident optical pulse must be $12Z$. Once again this neglects the Poisson statistics of the hole–electron pair production process. If the number of incident optical pulses per second (baud) is B and if hf is the energy in a photon, then for half "marks" (pulse present) and half spaces (pulse absent) the detected (taking into account quantum efficiency) optical power required for a 10^{-9} error rate is given by

$$P_{\text{av detected}} = 6hf\,BZ \tag{3.4.9}$$

Comparing to the quantum limit equation (3.4.5) we see that the required detected optical power is larger, due to amplifier noise, by the ratio $(6/10.5)Z$. Thus for $Z = 35{,}000$ we require about 43 dB more power than the quantum limit. To reduce this penalty one can take two approaches. One can construct amplifiers with lower values of Z; and one can use detectors with avalanche gain (to get more than one hole–electron pair per photon).

In order to reduce the amplifier thermal noise contribution one must take into account the fact that the photodetector is a "capacitive source".[1,3] That is, whereas many traditional signal sources are modeled as a current source in parallel with a resistor, as shown in Figure 3.16, the photodiode is modeled as a current source in parallel with a capacitor. Any parallel resistance associated with dark current or biasing resistors can usually be neglected for silicon detectors and for typical detector operating temperatures. (For longer-wavelength detectors, dark current becomes increasingly important.) The traditional resistive current source has a Johnson noise associated with it given by $4kT_{\text{eff}}/R_s$ ($A^2 \text{ Hz}^{-1}$), where T_{eff} is the effective noise temperature. The photodetector has no such noise source (except the shot noise of dark current which is typically negligible for silicon devices). In resistive source applications one can normalize the thermal noise contributed by the amplifier and the source combined to the thermal noise associated with the source alone, to obtain an amplifier noise figure. Since the capacitive source has essentially no thermal noise source, any such attempt to normalize the amplifier noise would typically result in an effective noise figure of essentially infinity, no matter how good the amplifier is. This is why in this chapter the amplifier figure of merit is defined via the parameter Z.

Capacitive sources are not unique to fiber optics. They appear in

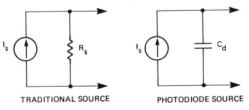

TRADITIONAL SOURCE   PHOTODIODE SOURCE

Figure 3.16. Signal sources.

vidicon applications, particle detector applications, and biological response measurement applications. The theory of design of low-noise amplifiers for optical detectors can therefore be adapted from these other applications.

A typical amplifier–detector combination is represented by the schematic shown in Figure 3.17. The amplifier is represented by a voltage-controlled current source g_m (siemens). The amplifier input impedance is the parallel combination of a capacitance C_a and a resistance R_a. There are two amplifier noise current sources which are Gaussian in statistics, flat in spectrum, and uncorrelated, given by N_1 and N_2 A² Hz⁻¹. Also shown is a feedback impedance Z_f. At the minimum Z_f represents parasitic capacitance between the amplifier output and input. There may also be an intentional feedback element. For the moment we shall neglect the feedback Z_f. We shall return to it in the discussion on transimpedance amplifiers which follows later.

One of the first observations, when Figure 3.17 is examined, is that the output of the current source $i_s(t)$ is in general producing a voltage across an an impedance whose value is not constant as a function of frequency. That is, the total impedance at the amplifier input is given by

$$Z_{\text{Total}} = \left[\frac{1}{R_a} + j2\pi f(C_a + C_d) \right]^{-1} \qquad (3.4.10)$$

There is a temptation to require a value of amplifier resistance R_a which is small enough so that the total impedance is controlled by R_a for all frequencies of interest. However, as we shall see, from a noise point of view, it is better to let R_a be relatively large. The roll-off at high frequencies which results at the amplifier input can be compensated for by equalization

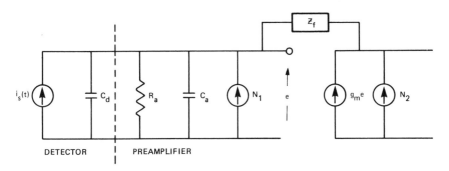

Figure 3.17. Typical amplifier–detector combination.

at the amplifier output. Such an amplifier–equalizer is shown in Figure 3.18. The receiver output signal is given by

$$V_{out}(f) = AI_s(f) \quad \text{for } f \le B$$
$$= 0 \quad \text{for } f > B \quad (3.4.11)$$

where A is an irrelevant amplifier gain, $I_s(f)$ is the Fourier transform of $i_s(t)$, $V_{out}(f)$ is the Fourier transform of $v_{out}(t)$, and B is the bandwidth of the band-limiting filter (assumed infinitely sharp for mathematical simplicity).

The noise at the receiver output is given by

$$\langle n_{out}^2 \rangle = N_1 A^2 B + N_2 A^2 B/(g_m R_a)^2 + (N_2 A^2 B^3/3)[2\pi(C_a + C_d)/g_m]^2 \quad (3.4.12)$$

We observe that in general there is a noise output term proportional to B and a term proportional to B^3. The total output noise depends upon N_1, N_2, R_a, C_a, C_d, and g_m. The gain A is irrelevant since it appears in both the rms noise and the signal given in (3.4.11).

Let us consider first the simple FET amplifier–equalizer shown in Figure 3.19. We assume that all thermal noise is produced in the preamplifier, and therefore we model the following amplifier stages as ideal. The equivalent circuit is shown in Figure 3.20. The resistance R_a represents the sum of detector and FET biasing resistors and is typically very large (a megohm or more). The noise source N_1 is the Johnson noise of these physical resistors: $4kT/R_a$ A^2 Hz^{-1}, which as we shall see is typically negligible. The noise source N_2 is the thermal noise of the channel between

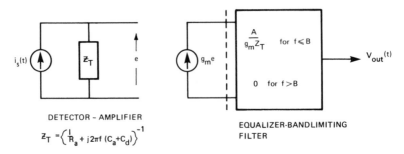

DETECTOR - AMPLIFIER

$$z_T = \left(\frac{1}{R_a} + j2\pi f \,(C_a + C_d) \right)^{-1}$$

EQUALIZER-BANDLIMITING FILTER

Figure 3.18. Amplifier equalizer.

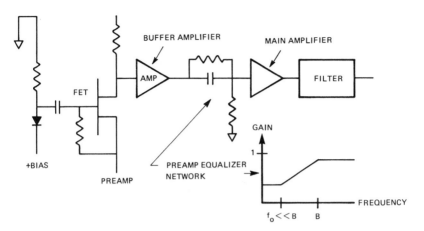

Figure 3.19. FET receiver.

source and drain, typically given by $2.8kTg_m$ A^2 Hz^{-1}. The amplifier–equalizer output noise is given by

$$\langle n_{out}^2 \rangle = 2.8kTg_m^{-1}\frac{B^3}{3}\left[2\pi(C_a + C_d)\right]^2 A^2 + \frac{4kTBA^2}{R_a} + \frac{2.8kTBA^2}{g_m R_a^2}$$

(3.4.13)

We see that for minimum noise R_a should be as large as possible. If we neglect the terms associated with R_a and its noise source $4kT/R_a$, we obtain

$$n_{out\ rms} = \left[\langle n_{out}^2 \rangle\right]^{1/2} = \left[2.8kTg_m^{-1}\frac{B^3}{3}\left[2\pi(C_a + C_d)\right]^2 A^2\right]^{1/2}$$

(3.4.14)

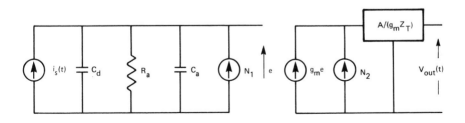

Figure 3.20. FET receiver equivalent circuit.

We observe that the noise at the output of the equalizer depends upon the parameter $(C_a + C_d)^2/g_m$. For a fixed material, say silicon, the ratio of C_a/g_m is fixed, although the values of C_a and g_m depend upon the FET geometry. Typically one uses available FET devices, but if one could optimize the FET to work with a given detector capacitance C_d, then by simple calculus one finds that for fixed C_a/g_m, $(C_a + C_d)^2/g_m$ is minimized with $C_a = C_d$. By using gallium arsenide FETs with a higher ratio of g_m/C_a, improved performance can be obtained. However, good GaAs FETs with low $1/f$ noise are difficult to fabricate relative to silicon FETs.

It is interesting to calculate the value of Z for a typical FET amplifier. Typical parameter values for a moderately good amplifier are $C_a + C_d = 10$ pF, $g_m = 5000$ μs, and R_a^{-1} negligibly small. If we set $B = 10$ MHz we obtain from (3.4.14)

$$n_{\text{out rms}} = 1.6 \times 10^{-9}A \qquad \text{(volts)} \qquad (3.4.15)$$

The area of the response at the equalizer output to an impulse of area e coulombs emitted by the current source $i_s(t)$ is given by

$$\text{area}_e = 1.6 \times 10^{-19}A \qquad \text{(volts)} \qquad (3.4.16)$$

The peak of the response at the equalizer output to a single generated hole–electron pair is this area multiplied by the amplifier bandwidth B (roughly). Thus Z is given by

$$Z = \frac{1.6 \times 10^{-9}A}{1.6 \times 10^{-19}A \times 10^7} = 1000 \qquad (3.4.17)$$

In general for the FET [compare to 3.4.7a)]

$$Z_{\text{FET}} = \left\{ \frac{2.8kTB^3}{3g_m} \left[2\pi(C_a + C_d) \right]^2 \right\}^{1/2} \bigg/ eB \qquad (3.4.17a)$$

Smaller values of Z can be obtained by reducing the total capacitance $C_a + C_d$. If the FET is optimized with $C_a = C_d$ and if C_a/g_m is fixed by the FET material, then the ratio Z is proportional to $(C_a + C_d)^{1/2}$. For this reason, in some applications it is desirable to build a hybrid circuit with a very low capacitance detector (around 1 pF) bonded with minimum lead capacitance to a substrate containing a silicon or GaAs FET of comparable capacitance.

From Figure 3.20 we see that the input to the FET amplifier is a current source in parallel with a capacitor. The output of the FET is also a current source in parallel with a capacitor. If we assume that the capacitances are roughly the same at input and output, it is interesting to ask whether we

have gained anything. The gain between the detector current source and
the FET voltage-controlled current generator is given by

$$\text{gain} = \frac{g_m E_{\text{in}}(f)}{I_s(f)} = f^{-1} g_m \left[2\pi (C_a + C_d) \right]^{-1} \qquad (3.4.18)$$

where g_m is the fixed FET transconductance. We see that at sufficiently
high frequencies, the gain approaches unity, and the FET ceases to perform
an amplifying function. A consequence of this is that at high frequencies
(about 25 MHz for silicon FETs) it is better to use a bipolar transistor
whose transconductance, g_m, is the ratio of its fixed current gain, β, to its
variable base input resistance. Figure 3.21 shows a typical bipolar amplifier–
equalizer. Figure 3.22 shows the equivalent circuit. The input resistance
of the transistor r_{in} is given by $4kT/(eI_{\text{base bias}})$, where $I_{\text{base bias}}$ is the base
bias current, under control of the amplifier designer. The parameter β is
the current gain of the transistor. R_a represents the parallel combination
of the transistor base bias resistors and the detector dc return resistor. The
input and output noise source spectral densities are given by

$$N_1 = \frac{4kT}{R_a} + 2eI_{\text{base bias}} = \frac{4kT}{R_a} + \frac{2kT}{R_{\text{in}}} \qquad (\text{amps}^2/\text{hertz}) \quad (3.4.19)$$

$$N_2 = 2e\beta I_{\text{base bias}} = 2eI_{\text{collector bias}} \qquad (\text{amps}^2/\text{hertz})$$

The signal and rms noise at the preamplifier output are given by (for
$R_a \gg r_{\text{in}}$)

$$V_{\text{out}}(f) = A I_s(f) \qquad \text{for } f \leq B$$

$$\langle n_{\text{out}}^2 \rangle = \frac{2kTA^2B}{r_{\text{in}}} + \frac{2kTA^2B}{\beta r_{\text{in}}} + \frac{2kTr_{\text{in}}A^2B^3}{3\beta} \left[2\pi(C_a + C_d) \right]^2$$

$$(3.4.20)$$

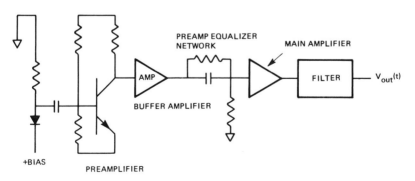

Figure 3.21. Bipolar receiver.

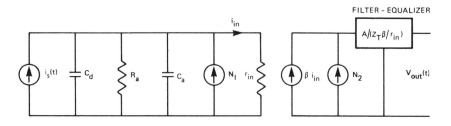

Figure 3.22. Bipolar receiver equivalent circuit.

We see that in the mean squared noise there is a term proportional to the bandwidth B and a term proportional to the cube of the bandwidth B^3. By varying the base bias current we can trade one term off against the other. The optimal value of r_{in} and optimized mean squared output noise are given by

$$r_{\text{in optimal}} = \left[2\pi(C_a + C_d)\,B\right]^{-1}(3\beta)^{1/2} \quad \text{(ohms)} \quad \text{for } \beta \gg 1$$

$$\langle n_{\text{out}}^2 \rangle_{\text{min}} = \frac{4kT\left[2\pi(C_a + C_d)\right]A^2B^2}{(3\beta)^{1/2}} = \frac{4kTBA^2}{r_{\text{in optimal}}} \quad \text{(volts}^2) \qquad (3.4.21)$$

We note that at optimal bias, the preamplifier mean squared output noise is proportional to B^2. We observe that $(C_a + C_d)^2/\beta$ is a figure of merit of amplifier noise performance. As with the FET, since β/C_a is fixed for a given material, an optimized amplifier would have $C_a = C_d$. We also observe that at optimal bias, the bandwidth of the input impedance is $B/(3\beta)^{1/2}$. Thus indeed the amplifier input integrates the signal, while the equalizer differentiates to restore the waveform.

The value of Z can be calculated as for the FET. It is given by

$$Z_{\text{bipolar}} = \left\{ \frac{4kT\left[2\pi(C_a + C_d)\right]}{(3\beta)^{1/2}} \right\}^{1/2} \left(\frac{1}{e}\right) = \left(\frac{4kTB}{r_{\text{in optimal}}}\right)^{1/2} \Big/ eB \qquad (3.4.22)$$

Note that for an optimized bipolar amplifier, Z is independent of B. [Using the first expression in (3.4.22). Remember, $r_{\text{in optimal}}$ is inversely proportional to B.]

The parameter Z is also $(3\beta)^{1/4}$ times smaller than it would be if a physical resistor of value $R = [2\pi B(C_a + C_d)]^{-1}$ were placed across the amplifier input to prevent integration (see 3.4.7a). If we set $\beta = 100$, and $C_a + C_d = 10$ pF, we obtain $Z = 1500$.

Another interesting parameter associated with amplifiers for capacitive sources is the "noise corner frequency." If we look at the noise in a differential bandwidth, df, as a function of frequency at the equalizer output we obtain a curve as shown in Figure 3.23. The frequency f_c is the noise corner frequency. For an FET as shown in Figures 3.19 and 3.20 the parameter f_c is obtained from (3.4.13) (for $R_a g_m \gg 1$) by differentiating with respect to B:

$$N_{\text{out}}(f) = \frac{2.8kTf^2}{g_m} A^2 [2\pi(C_a + C_d)]^2 + \frac{4kT}{R_a} A^2 \quad \text{(volts}^2/\text{hertz)}$$

$$(3.4.23)$$

The noise corner occurs at

$$f_c = [2\pi(C_a + C_d) R_a]^{-1} [(4/2.8) g_m R_a]^{1/2} \quad \text{(hertz)}$$

For a bipolar amplifier shown in Figures 3.21 and 3.22, f_c is given by (for $\beta \gg 1$)

$$f_c = [2\pi(C_a + C_d) r_{\text{in}}]^{-1} \beta^{1/2} \quad (3.4.24)$$

for an optimized bipolar amplifier $f_c = B/3^{1/2}$.

It is also interesting to compare the output noise of a typical FET and a typical bipolar amplifier. Figure 3.24 shows the parameter Z as a function of bandwidth for the amplifiers considered above. Once again we see that for an optimized bipolar amplifier, Z is constant with bandwidth while for the FET case Z varies as $B^{1/2}$. The frequency $f_{\text{max FET}}$ is where the signal

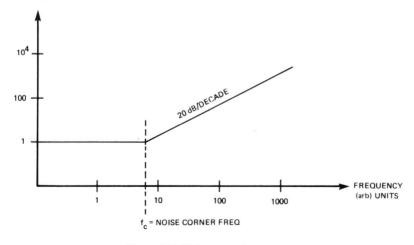

Figure 3.23. Noise corner frequency.

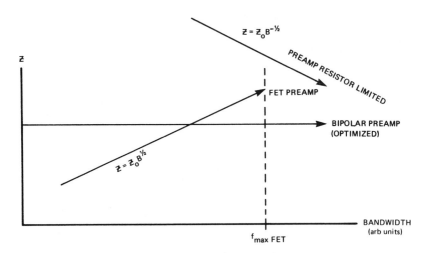

Figure 3.24. Z vs. bandwidth for various amplifiers.

gain of the FET equals unity. The implication of this is, of course, that if the FET gain is near unity, one cannot neglect noise added by subsequent amplifier stages.

In many applications, noise performance is not the only major consideration. Often dynamic range is an equally important parameter. Receivers must be designed not only to work with the weakest allowable light signals, but they must also accommodate larger light signals without overload.

One problem with the integrate–differentiate type of amplifier described above is that before differentiation it has larger gain at low frequencies, which makes it prone to overload with signals having large low-frequency content. If the amplifier overloads for whatever reason, the integrated waveform cannot be restored by differentiation.

An alternative design is the transimpedance amplifier shown in Figure 3.25.[9] Typically one assumes that in such an amplifier the loop gain is sufficiently large so that the relationship between the amplifier output voltage and the detector output current is given by

$$V_{\text{out}}(f) = Z_f I_s(f)$$

provided

$$A_f Z_{\text{in}}/(Z_{\text{in}} + Z_f) \gg 1 \qquad (3.4.25)$$

In order to compare the transimpedance amplifier to the integrate–differentiate types discussed above, consider the example of a transimpedance

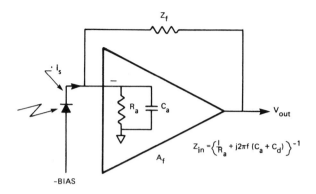

Figure 3.25. Transimpedance amplifier.

amplifier shown in Figure 3.26. The preamplifier is the same as the integrating type considered above, and we assume that, except for the Johnson noise of the feedback element, the preamplifier is the dominant noise source. The amplifier Z_2 provides additional loop gain and a low output impedance. The noise source N_f is the Johnson noise of the feedback element. The equalizer is included for generality and to make the overall amplifier–equalizer gain the same as for the amplifier–equalizers considered above. That is,

$$V_{\text{out}}(f) = A\, I_s(f) \qquad \text{for } f \le B$$
$$= 0 \qquad\qquad \text{otherwise} \qquad (3.4.26)$$

The noise at the equalizer output is given by

$$\langle n_{\text{out}}^2 \rangle = \int_0^B \left\{ N_1 A^2 + N_f A^2 + \frac{N_2 A^2 \gamma}{g_m^2 R_a^2} + \frac{N_2 A^2 f^2}{3 g_m^2} \left[2\pi(C_a + C_d) \right]^2 \gamma \right\} df \qquad (3.4.27)$$

where

$$\gamma = \left| \frac{Z_{\text{in}} + Z_f}{Z_f} \right|^2 \quad \text{and } Z_{\text{in}} = \left[\frac{1}{R_a} + j2\pi f(C_a + C_d) \right]^{-1}$$

Comparing (3.4.27) to (3.4.12) we see that if Z_f/Z_{in} is much greater than unity (as is typical), and therefore $\gamma \sim 1$, then the output noise of the transimpedance amplifier is the sum of the output noise of an integrate–differ-

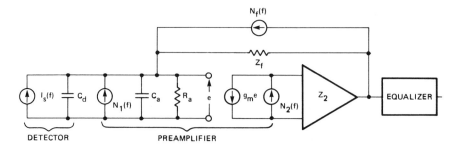

Figure 3.26. Transimpedance amplifier (detail).

entiate amplifier plus the noise associated with the feedback element. In other words,

$$Z^2_{\text{transimpedance}} = Z^2_{\text{FET or bipolar}} + 4kTG_fB/(eB)^2 \qquad (3.4.28)$$

where G_f is the real part of $(Z_f)^{-1}$.

In many designs, Z_f is a resistor R_f (plus stray capacitance). For a bipolar transistor amplifier as described above we then have the relationship

$$Z^2_{\text{transimpedance}} = \left(\frac{4kTB}{r_{\text{in optimal}}} \Big/ e^2B^2 \right) + \left(\frac{4kTB}{R_f} \Big/ e^2B^2 \right) \qquad (3.4.29)$$

Thus the noise of the feedback resistance R_f given by (3.4.29) adds to the noise of the equivalent integrate–differentiate amplifier input noise resistance $r_{\text{in optimal}}$. Typically to allow for a reasonable dynamic range R_f is often smaller than $r_{\text{in optimal}}$ and therefore its noise contribution dominates. If this is the case, the preamplifier biasing may be adjusted for increased dynamic range, since the noise is controlled by the feedback resistor. The value of the feedback resistor is limited not only by dynamic range requirements, but also by the requirement for adequate loop gain around the feedback path in typical operation.

We shall give an example of an actual transimpedance amplifier used in an optical fiber system in Section 3.4.6 after the section on avalanche detectors.

3.4.3. Avalanche Gain [10-13]

In addition to requiring low-noise amplifiers, the most sensitive optical receivers incorporate avalanche detectors. In such a device, each primary hole–electron pair created by optical absorption generates tens or hundreds of secondary hole–electron pairs through the process of collision ionization. Thus each primary hole–electron pair is replaced by a "bunch" of secondary pairs. Ideally, the number of secondary pairs generated by a primary pair would be completely deterministic. Thus the displacement current which would flow in the detector when a primary pair is generated would have area Me coulombs, where M is the multiplication or gain of the APD and e is the electron charge. This would have an effect on receiver sensitivity equivalent to reducing the receiver noise parameter Z by the factor M. Unfortunately the multiplication process is statistical. That is, each primary hole–electron pair generates a random number of secondary pairs because of the statistical nature of the multiplication process.

Consider first an example of how statistical multiplication affects the receiver performance. Suppose the APD is illuminated by a pulse of light having energy $E = \Lambda hf$, where Λ is the average number of detected photons and hf is the energy in a photon. The number of primary hole–electron pairs generated is governed by the Poisson detection statistics (quantum noise) and has probability distribution

$$p(n) = \text{probability of } n \text{ generated primary hole–electron pairs}$$

$$= \Lambda^n e^{-\Lambda}/n! \tag{3.4.30}$$

The mean and standard deviation of the number of hole–electron pairs generated is given by

$$n_{\text{av}} = \Lambda$$

$$\sigma_n = \Lambda^{1/2} \tag{3.4.31}$$

Now assume that the detector gain is governed by the distribution $p_M(m) = $ probability that a primary hole electron pair produces m secondary pairs including the primary. We then obtain the total distribution of secondary pairs by averaging $p_M(m)$ over the statistics of the primary generation process:

$$p_{\text{tot}}(n) = \text{probability of } n \text{ secondary pairs}$$

$$= \sum_{k=0}^{\infty} p_M^{*k}(m)\, \Lambda^k e^{-\Lambda}/k! \tag{3.4.32}$$

where

$$p^{*k} = p \text{ convolved with itself } k \text{ times}$$

$$= p * p * p \cdots \quad (k \text{ times})$$

and where

$$p_1 * p_2 = \sum_{k=-\infty}^{\infty} p_1(n-k)p_2(k)$$

The mean and variance of the total number of secondary pairs are given by (no approximations)

$$n_{\text{av tot}} = \Lambda M$$

$$\sigma_{\text{tot}}^2 = \Lambda(M^2 + \sigma_M^2) = \Lambda \langle m^2 \rangle \tag{3.4.33}$$

where M is the mean of the avalanche multiplication and σ_M is the standard derivation of the multiplication.

For an ideal multiplication mechanism with exactly M secondary pairs per primary, σ_M would be zero. The mean squared gain $<m^2>$ is a rough measure of the effect of the unpredictability of the avalanche mechanism on the receiver performance. More quantitative results will be given in Section 3.4.4. Very often avalanche detectors are characterized by an "excess noise factor" F_M defined as

$$F_M = \langle m^2 \rangle / M^2 = 1 + (\sigma_M^2/M^2) \tag{3.4.34}$$

Before presenting formulas for typical values of F_M vs. M, let us consider in some detail the statistics of the simplest type of avalanche detector (and also the best from a statistical point of view).

This APD is one in which only one type of carrier, say electrons, can generate hole–electron pairs by the collision ionization process. The detector and the multiplication process are depicted in Figure 3.27. An electron moving toward the left in the high-field region has a probability αdx of generating a hole–electron pair as it moves an incremental distance dx. Let distance x be measured right to left as shown. Let $P_{M,x}(m)$ be the probability distribution of the total number of electrons M at position x generated as a result of the initial primary electron entering from the right. Each electron at position x has a probability of generating an additional electron (and hole) as it moves an increment dx toward the left. Thus neglecting terms of higher order than dx we have

$$p_{M,x+dx}(m) = p_{M,x}(m)(1 - m\alpha dx) + p_{M,x}(m-1)\alpha[m-1]dx \tag{3.4.35}$$

That is, to have m electrons at $x + dx$, either there are m electrons at x and no more generated in dx, or there are $m - 1$ electrons at x and one additional electron is generated in dx. We obtain

$$\frac{\partial}{\partial x}p_{M,x}(m) = (m-1)\alpha p_{M,x}(m-1) - m\alpha p_M(x) \tag{3.4.36}$$

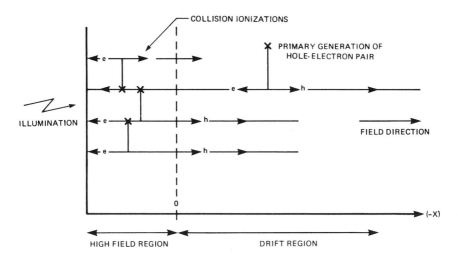

Figure 3.27. Avalanche multiplication (single-carrier ionizing).

The solution of (3.4.36) is given by [allowing $\alpha(x)$ to vary with x]

$$P_{M,x}(m) = \frac{1}{M(x) - 1} \left[\frac{M(x) - 1}{M(x)} \right]^m \quad \text{for } m \geq 1 \quad (3.4.37)$$

where

$$\delta(x) = \int_0^x \alpha(x')dx'$$

$$M(x) = \text{average gain} = e^{\delta(x)}$$

$$\sigma_M^2 = M^2(x) - M(x)$$

$$F_M = \left[2 - \frac{1}{M(x)} \right]$$

The distribution (3.4.37) is known as a Bose–Einstein distribution. It is characterized by the single parameter $M(x)$, the mean multiplication, which is related to $\alpha(x)$ as given above. The excess noise factor, F_M, converges to 2 for large gains $M(x)$. The distribution is shown in Figure 3.28 along with the ideal deterministic distribution. The fact this distribution is (approximately) exponential in shape (if we smooth out the discreteness)— and since as stated this is one of the better distributions from a receiver performance point of view—should give the reader some insight into how noncompact avalanche gain distributions are.

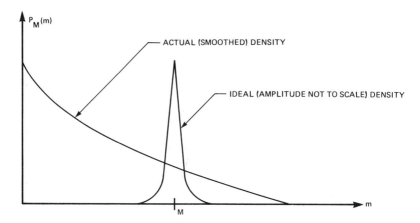

Figure 3.28. Actual and ideal gain densities.

In a more general avalanche detector, both holes and electrons can produce secondary hole–electron pairs. Such detectors are characterized by the carrier ionization ratio k which is defined as the ratio of the ionization probabilities per unit length (in the high-field region) of holes and electrons. Ideally for good statistics the avalanche should be initiated by the more strongly ionizing carrier. That is, the primary hole–electron pair generated by light absorption should be generated outside of the high-field region, with the more strongly ionizing carrier drifting or diffusing into the region. For a detector with a mean avalanche gain M and an ionization ratio k, the probability distribution $p_M(m)$ of the multiplication has been determined to satisfy the following equation:

$$p_M(m) = (1 - k)^{m-1} \, \Gamma\left(\frac{m}{1-k}\right) \left[\frac{1 + k(M-1)}{M}\right]^{[1+k(m-1)]/(1-k)}$$

$$\left(\frac{M-1}{M}\right)^{m-1} \left\{ [1 + k(m-1)](m-1)! \, \Gamma\left(\frac{1 + k(m-1)}{1-k}\right) \right\}^{-1}$$

$$(3.4.38)$$

$$F_M = \left[kM + \left(2 - \frac{1}{M}\right)(1 - k) \right]$$

where $\Gamma(x)$ is the gamma function. Figure (3.29) shows curves of F_M vs. M with k as a parameter.

Various approximations for $p_M(m)$ given in (3.4.38) have been derived for use in performance calculations. These will be discussed in Section 3.4.4.

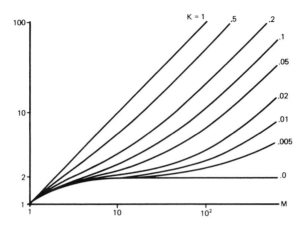

Figure 3.29. F_M vs. M, $F_M = kM + (2-1/M)(1-k)$.

3.4.4. Performance Calculations for Binary Digital Receivers [1, 14-16]

In this section we shall calculate the error rate of a binary digital receiver taking into account the statistics of photodetection, avalanche detection statistics, amplifier thermal noise, dark current, and certain nonidealities of the received signal including pulse overlap and imperfect extinction ratios. A typical binary receiver with a fixed pulse rate (clocked) is shown in Figure 3.30. Optical pulses arrive at a fixed rate B (baud). Each pulse has a basic shape $h_p(t)$ where $h_p(t)$ is normalized to area unity.

$$\int h_p(t)\, dt = 1 \qquad (3.4.39)$$

If in a given time a slot pulse is "on," then it has energy E_r or if it is "off," it has energy $\mathrm{EXT} \cdot E_r$. The received light pulses fall upon an APD or a pin detector to produce a photocurrent. The detector is assumed to have a sufficiently fast response so that the average photocurrent is a replica of the received optical power. The photocurrent is corrupted by the statistical uncertainties of photodetection and avalanche gain (if present). The preamplifier adds thermal noise to the detected signal. The first equalizer compensates for any roll-offs in the preamplifier. The second equalizer band-limits and shapes the responses to individual light pulses so that the output of the second equalizer is given by

$$V_{\mathrm{out}}(t) = A \sum_{k=-\infty}^{\infty} a_k h_{\mathrm{out}}(t - kT) + \mathrm{noise} \qquad (3.4.40)$$

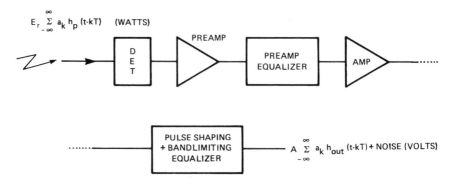

Figure 3.30. Typical binary digital receiver.

where A is an irrelevant constant

$$a_k = 1 \text{ or EXT}$$

$$T = 1/B = \text{pulse "time slot" width}$$

$$B = \text{baud}$$

and where $h_{\text{out}}(t)$ is the basic output pulse shape produced when an optical pulse of shape $h_p(t)$ is incident upon the receiver. The voltage waveform $v_{\text{out}}(t)$ is sampled once per pulse interval to determine whether or not a pulse is present. If $v_{\text{out}}(t)$ exceeds a threshold, γ, a pulse is assumed present and vice versa. In order to calculate the probability of making a decision error, one needs to know the statistics of the voltage $v_{\text{out}}(t)$ at the sampling time. These statistics are very complicated, and various tradeoffs have been made between the accuracy of the calculation and computational simplicity.

The simplest approximation to calculation of the error rate is called the Gaussian approximation.[1] In this approximation one assumes that given the input optical pulse sequence (which pulses are on and which are off) the output voltage $v_{\text{out}}(t)$ is a Gaussian random variable. Within this approximation it is only necessary to know the mean and standard deviation of $v_{\text{out}}(t)$ in order to calculate the error probability. An example of such a calculation follows.

For simplicity assume that individual received pulses $h_p(t)$ do not overlap, and that at the sampling time jT, the output voltage $v_{\text{out}}(t)$ is proportional to the number of hole–electron pairs produced by the photodetector in response to optical pulse j, plus thermal noise from the preamplifier. That is

$$v_{\text{out}}(jT) = K_1 \left(\sum_{l=1}^{N_j} m_l + n_{\text{th}} \right) \qquad (3.4.41)$$

where K_1 is an irrelevant constant, m_l is the number of secondary hole-electron pairs produced by primary pair l, N_j is the number of primary hole–electron pairs generated by pulse j, and n_{th} is the preamplifier thermal noise with standard deviation Z.

The mean and standard deviation of v_{out} if the received pulse is "off" are given by

$$\langle v_{out\ "off\ "}\rangle = K_1\left(\frac{EXT \cdot E_r}{hf} \cdot M\right) \qquad (3.4.42)$$

where $EXT \cdot E_r$ is the received optical pulse energy in the "off" state, hf is the energy in a photon, M is the mean avalanche gain, and

$$\sigma^2_{"off\ "} = K_1^2\left(\frac{EXT \cdot E_r}{hf}M^2F_M + Z^2\right)$$

where F_M is the avalanche detector excess noise factor and Z is the preamplifier noise parameter.

If we wish to keep the error probability below 10^{-9}, then for the assumed Gaussian statistics of v_{out} we require the threshold to be six standard deviations above the average value of v_{out} in the "off" state. This will assure (for Gaussian statistics) that only one time out of a billion will the voltage v_{out} exceed the threshold if the received optical pulse is "off." Thus we require

$$\gamma = \langle v_{out\ "off\ "}\rangle + 6\sigma_{"off\ "} \qquad (3.4.43)$$

where γ is the threshold for decisions.

If the optical pulse is "on," the mean and standard deviations of v_{out} are similarly given by

$$\langle v_{out\ "on\ "}\rangle = K_1\left(\frac{E_r}{hf} \cdot M\right) \qquad (3.4.44)$$

$$\sigma^2_{"on\ "} = K_1^2\left(\frac{E_r}{hf}M^2F_M + Z^2\right)$$

To maintain a 10^{-9} error rate, the average signal at the sampling time in the "on" state must exceed the threshold by six standard deviations of the value of $v_{out}(t)$ in the "on" state

$$\langle v_{out\ "on\ "}(t)\rangle = \gamma + 6\sigma_{"on\ "} \qquad (3.4.45)$$

Combining (3.4.42)–(3.4.45) we obtain the requirement on E_r/hf:

$$\frac{E_r}{hf} \cdot M[1 - \text{EXT}] = 6\left(\frac{E_r}{hf}M^2 F_M + Z^2\right)^{1/2}$$

$$+ 6\left(\frac{E_r}{hf}M^2 F_M \cdot \text{EXT} + Z^2\right)^{1/2} \qquad (3.4.46)$$

For the case EXT $= 0$ (perfect extinction in the "off" state) it is relatively easy to solve (3.4.46) to obtain

$$\frac{E_r}{hf} = 36F_M + 12\frac{Z}{M} \qquad (3.4.47)$$

Equation (3.4.47) includes the Gaussian approximation, the perfect extinction approximation, the "no interference from adjacent optical pulses" approximation, and the "output pulse proportional to the total number of secondary hole–electron pairs" (no weighting due to filter impulse response shapes) approximation. We can make several observations from (3.4.47). If there is no amplifier noise ($Z = 0$) and if we therefore set the avalanche gain M to unity and $F_M =$ unity, then the equation says that the required number of photons in an "on" optical pulse is 36. The number calculated from the exact Poisson statistics is 21 (for a 10^{-9} error rate) as obtained in Section 3.4.1 above. The discrepancy of 36/21 is due to the Gaussian approximation. If we assume Z is large compared to unity and if we set M and F_M equal to unity (pin detector), then we find that the required number of photons in an "on" pulse is $12Z$ as described in Section 3.4.2 above. Since thermal noise has Gaussian statistics, this is a very accurate result. We also see that if Z is much larger than unity (typical) the required number of photons per optical pulse scales as $1/M$ until the term involving F_M dominates. Figure 3.31 shows curves of the required number of photons per optical pulse (in the "on" state for a 10^{-9} error rate) vs. M, for $Z = 1000$ and various values of detector parameter k [see equation (3.4.38)]. We see that there is an optimal value of avalanche gain M approximately equal to 50–100 for digital receivers, for $Z = 1000$, $k \cong 0.02$–0.10 and with a 10^{-9} error rate. The resulting value of E_r/hf for an "on" pulse is around 400–800, or about 13–16 dB more than the quantum limit of 21.

The above approximations can be removed at the expense of increasing computational complexity. Although we shall leave the details to the references,[14] there are two aspects of the approximations worth discussing further: the Gaussian approximation and the consequences of intersymbol interference.

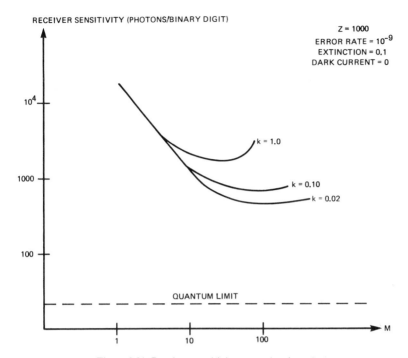

RECEIVER SENSITIVITY (PHOTONS/BINARY DIGIT)

Z = 1000
ERROR RATE = 10^{-9}
EXTINCTION = 0.1
DARK CURRENT = 0

k = 1.0
k = 0.10
k = 0.02

QUANTUM LIMIT

Figure 3.31. Receiver sensitivity vs. avalanche gain.

The advantage of the Gaussian approximation is that one requires only the mean and standard deviations of the receiver output voltage, and further that these quantities are fairly easy to obtain, even in much more general situations than considered above. The exact statistics of the output voltage at the sampling time are very difficult to derive in general but can be obtained numerically in limited situations. One can bootstrap one's confidence in the Gaussian and other approximations by comparing results of approximate and exact methods in those special cases where both are obtainable. For example, we derived above that in the quantum limit, of no thermal noise from amplifiers, the exact Poisson statistics of photodetection predicted 21 photons required for a 10^{-9} error rate, whereas the Gaussian approximation predicted 36 photons. Thus in this special case the Gaussian approximation is 1.5 dB pessimistic.

One known problem with the Gaussian approximation is the fact that it does not take into account the extreme skewness of the avalanche gain distribution. In particular it underestimates the probability that, in the "off" state, a small number of primary hole–electron pairs (generated by imperfect extinction or dark current) can produce a very large number of secondary pairs, thus causing an incorrect threshold crossing (error). An alternative to the Gaussian approximation which is analytically and numerically tractable and which accounts more for the details of the gain statistics is the Chernoff bound.[8]

The Chernoff bound is based on the moment-generating function of a probability distribution. That is, if $P_X(x)$ is a probability density, then $M_X(s)$, the moment-generating function, is defined as

$$M_X(s) = \int_{-\infty}^{\infty} p_X(x) e^{sx} dx \qquad (3.4.48)$$

The Chernoff bound for the probability that random variable X will exceed a threshold γ is derived as follows:

$$\text{Prob}(x \geq \gamma) = \int_{\gamma}^{\infty} p_X(x) dx \leq \int_{\gamma}^{\infty} p_X(x) e^{s(x-\gamma)} dx \qquad \text{for } s \geq 0 \quad (3.4.49)$$

Therefore for $s \geq 0$

$$\text{Prob}(x \geq \gamma) \leq \int_{-\infty}^{\infty} p_X(x) e^{sx} dx \, e^{-s\gamma} = M_X(s) e^{-s\gamma}$$

Defining the characteristic function

$$\psi_X(s) = \ln M_X(s) \qquad (3.4.50)$$

we obtain

$$\text{Prob}(x \geq \gamma) \leq e^{\psi_X(s) - s\gamma} \qquad \text{for } s \geq 0 \qquad (3.4.51)$$

Differentiating the right-hand side of (3.4.51) to obtain the tightest bound, we obtain

$$\text{Prob}(x \geq \gamma) \leq e^{\psi_X(s) - s\psi'_X(s)}\big|_{\psi'_X(s) = \gamma} \qquad \text{for } s \geq 0 \qquad (3.4.52)$$

where $\psi'(s) = (\partial/\partial s)\psi(s)$. Similarly we obtain

$$\text{Prob}(x \leq \gamma) \leq e^{\psi_X(s) - s\psi'_X(s)}\big|_{\psi'_X(s) = \gamma} \qquad \text{for } s \leq 0 \qquad (3.4.52a)$$

In order to understand the Chernoff bound, let us apply it to a quantum-limited detection problem. Specifically, given that a pulse of energy $E_r = \Lambda hf$ arrives at an ideal photodetector, what is the probability that less than two hole–electron pairs are produced? The number of hole–electron pairs produced is governed by Poisson statistics: $p_N(n) = \Lambda^n e^{-\Lambda}/n!$. From (3.4.52) we obtain

$$\text{Prob}(n < 2) = \text{Prob}(n < \gamma) \leq e^{\psi_N(s) - s\psi'_N(s)}\big|_{\psi'(s) = \gamma} \qquad (3.4.53)$$

for $s \leq 0$ and any threshold, γ, such that $1 < \gamma < 2$ (since n is quantized to multiples of unity). We have

$$M_N(s) = \sum_{n=0}^{\infty} e^{sn} \frac{\Lambda^n e^{-\Lambda}}{n!} = \sum_{n=0}^{\infty} \frac{(\Lambda e^s)^n e^{-(\Lambda e^s)}}{n!} [e^{\Lambda(e^s-1)}] = e^{\Lambda(e^s-1)} \quad (3.4.54)$$

$$\psi_N(s) = \Lambda(e^s - 1)$$

$$\psi_N'(s) = \Lambda e^s$$

Thus

$$\text{Prob}(n < 2) \leq e^{\Lambda(e^s - 1 - se^s)}\big|_{\Lambda e^s = \gamma} = e^{-\Lambda} e^{\gamma} \left(\frac{\Lambda}{\gamma}\right)^{\gamma}$$

$$= (\text{for } \gamma \to 1) \, e^{-\Lambda} \Lambda e$$

provided $s < 0$, which means $\Lambda > 1$.

The exact result is

$$\text{Prob}(n < 2) = e^{-\Lambda}(1 + \Lambda)$$

$$= e^{-(\Lambda+1)}(\Lambda + 1) e \quad (3.4.55)$$

Thus in this special case, the Chernoff bound is conservative in predicting probability by the factor $e/(1 + \Lambda^{-1})$. For an probability of 10^{-9} the Chernoff bound predicts a required value of Λ of 24.9, compared to the exact result of $\Lambda = 23.9$. Thus in this case, the Chernoff bound is conservative by 0.17 dB.

The advantage of the Chernoff bound is that the characteristic function of the random variable $v_{out}(t)$ at the sampling time is derivable and amenable to numerical error rate calculations. For example consider the special situation described above and summarized in equation (3.4.41). If an optical pulse is present having mean number of photons E_r/hf, then the characteristic function of $v_{out}(jT)$ is given exactly by

$$M_v(s) = (E_r/hf) [M_M(sK_1) - 1] + s^2 K_1^2 Z^2 / 2 \quad (3.4.56)$$

where $M_M(s)$ is the moment-generating function of the avalanche gain distribution. Differential equations for $M_M(s)$ have been derived. In particular for a specified detector parameter k and mean gain M one has the equation

$$\frac{\partial}{\partial s} M_M(s) = M_M(s) \frac{k-1}{k} \left\{ 1 - \frac{1}{k} [M_M(s) e^{-s} e^{\delta}]^{k-1} \right\}^{-1} \quad (3.4.57)$$

where

$$M = \frac{k-1}{k}\left(1 - \frac{1}{k}e^{\delta(k-1)} \right)^{-1} \qquad (3.4.57)$$

Using (3.4.56), its equivalent for the optical pulse "off" situation, and using (3.4.57) one can obtain a bound on the required optical energy E_r to obtain desired error rate. (Results to be presented below.)

In order to calibrate the accuracy of the Chernoff bound (or any other approximation) it is helpful to obtain exact error rate calculations. We have already done so at the quantum limit extreme. It is possible to use the exact equations for avalanche gain statistics and the Poisson statistics of the detection process to obtain other comparison points which include the effects of thermal noise and avalanche gain. This can be done using brute force approaches on a computer, or with more elegant techniques such as Monte Carlo type approaches.

Figure 3.32 shows a comparison of calculated values of the required number of photons per pulse E_r/hf vs. the avalanche gain setting using the Gaussian approximation, the Chernoff bound, and exact methods. The parameter values used in this figure are avalanche detector parameter $k = 0.1$, $Z = 6000$, EXT = 0.01, 5 dark current pairs per time slot, and an error rate of 10^{-9}. Figure 3.32 also shows the threshold position.

We observe that the Gaussian approximation and the Chernoff bound are in fairly good agreement in predicting the required number of photons

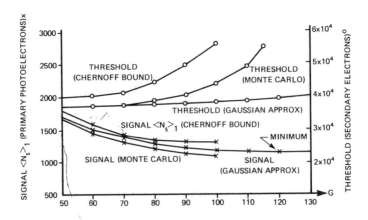

Figure 3.32. Comparison of gaussian approximation, Chernoff bound, and exact calculations of receiver sensitivity [from Personick *et al.*, 1977, *IEEE Trans. Commun. COM-25*, 546 Figure 4].

in the "on" state. The Gaussian approximation overestimates the optimal avalanche gain and also underestimates the optimal threshold level. This confirms the fact that the Gaussian approximation underestimates the effect of the skewness in the avalanche gain distribution on the probability of threshold crossings from below in the "off" state.

In order to calculate error rates, various approximations to the avalanche gain distribution have been derived and compared numerically to exact results. One excellent approximation for the total number of secondary pairs produced when the number of primary pairs is Poisson distributed with mean Λ is [see equation (3.4.32)]

$$p_{tot}(n) = \text{Prob}\left(\sum_{k=1}^{N} m_k = n \right) = \sum_{N=0}^{\infty} p_M^{*N}(n) \Lambda^N e^{-\Lambda}/N!$$

$$\cong \frac{1}{(2\pi)^{1/2}\,\sigma_n} \frac{1}{[1 + (n - \bar{n})/\sigma_n\lambda]^{3/2}} \exp\left\{ \frac{-(n - \bar{n})^2}{2\sigma_n^2[1 + (n - \bar{n})/\sigma_n\lambda]} \right\}$$

(3.4.58)

where

$$\bar{n} = \text{average value of } n = \Lambda M$$

$$\sigma_n^2 = \Lambda M^2 F_M, \qquad F_M = kM + (2 - 1/M)(1 - k),$$

$$\lambda = [\Lambda F_M/(F_M - 1)]^{1/2}$$

and where

$$p_M(n) \cong \frac{1}{(2\pi)^{1/2}} \frac{M^{1/2} F_M (1/n)^{3/2}}{(F_M - 1)^{3/4}} \exp\{-n/[2M(F_M - 1)^{1/2}]\}$$

Note that the tail of the exact avalanche gain distribution also decreases exponentially. For this reason, the moment-generating function, $M_M(s)$ [see equation (3.4.57)], "blows up" for values of s exceeding s_{crit} given by

$$s_{crit} = -\ln\left[\left(1 - \frac{1}{M} \right)\left(1 + \frac{1 - k}{kM} \right)^{k/(1-k)} \right]$$

(3.4.59)

Calculations made with a computer using the moment-generating function must be capable of handling this singularity. Figure 3.33 shows the typical behavior of the moment-generating function. Note that the derivative of $M_M(s)$ with respect to s goes to infinity at s_{crit} whereas $M_M(s)$ is still finite.

For more details on various approaches to calculation of the error rate, the reader is referred to the references.

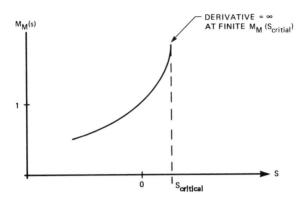

Figure 3.33. Typical behavior of moment-generating function of avalanche gain.

In the above discussions we have neglected the effects of "intersymbol interference." More specifically, we assumed that the received optical pulses having shape $h_p(t)$ did not overlap. In addition, we assumed that at the sampling time, the receiver output $v_{out}(jT)$ was proportional to the number of hole–electron pairs produced by optical pulse j. We excluded the possibility that hole–electron pairs from other optical pulses might contribute to the receiver output at time jT. In general several constraints work against these assumptions. The received optical pulses can overlap if pulse spreading has occurred as a result of transmission of the transmitter output pulses through the fiber (delay distortion). In addition, when the receiver is designed one tries to limit the bandwidth to minimize noise. This causes neighboring pulses to contribute noise and interference at the sampling time for a given pulse j. We shall discuss the interactions of transmitted pulse shape, fiber delay distortion, and receiver design in Chapter 4. However, we can make some general observations in this section.

All other things being equal, for a given received energy E_r in an "on" pulse, the optimal pulse shape $h_p(t)$ would be very narrow compared to a time slot $T = 1/B$. The receiver sensitivity (that is, the value of E_r required to achieve a desired error rate) would be optimized for such a pulse shape. The actual receiver sensitivity differs from this optimal value by an amount depending upon the specific receiver design and the specific pulse shape $h_p(t)$. One can derive some fairly representative curves illustrating the "receiver sensitivity penalty" associated with a received pulse shape $h_p(t)$ that deviates from the ideal. Before presenting such curves, it is useful to define a parameter σ_r called the "rms pulse width" of the received optical pulse. Since the received optical pulse represents power, it is, of course, a positive quantity. We have already normalized $h_p(t)$, the pulse shape, to

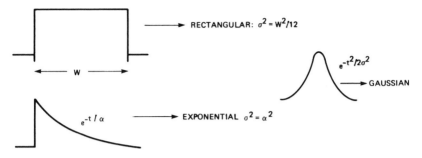

Figure 3.34. Received pulse shapes.

have area unity. The rms width of $h_p(t)$ is essentially the standard deviation of $h_p(t)$, in time, from its center of gravity. That is,

$$\sigma_r^2 = \int_{-\infty}^{\infty} (t - \bar{t})^2 \, h_p(t) \, dt \qquad (3.4.60)$$

where

$$\int h_p(t) \, dt = 1$$

and

$$\bar{t} = \int t h_p(t) \, dt$$

Figure 3.34 shows three types of received pulse shapes: rectangular, Gaussian, and exponential. Figure 3.35 shows the receiver sensitivity penalty in dB (relative to $\sigma_r B = 0$) for a typical receiver incorporating a

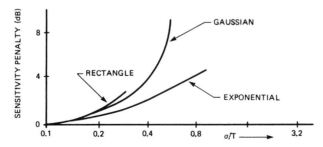

Figure 3.35. Receiver sensitivity penalty vs. σ/T PIN detector (from *Bell Syst. Tech. J.*, July–Aug. 1973, p. 867, Figure 19, copyright 1973 American Telephone and Telegraph Company).

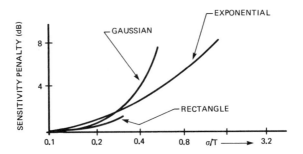

Figure 3.36. Receiver sensitivity penalty vs. σ/T APD detector (from *Bell Syst. Tech. J.*, July–Aug. 1973, p. 868, Figure 20, copyright 1973 American Telephone and Telegraph Company).

PIN detector as a function of $\sigma_r B = \sigma_r/T$. Figure 3.36 shows similar curves for an APD receiver. These curves were derived using the Gaussian statistical approximation for $v_{out}(t)$. We observe that for σ_r/T less than 0.25 the receiver sensitivity penalty is less than 1 dB.

In these calculations, it was assumed that the receiver would compensate for the received pulse shape $h_p(t)$ by "equalizing" to a desired output pulse shape $h_{out}(t)$ as described in equation (3.4.40) above. Thus the receiver has to "know" the received pulse shape. Another approach is to design the receiver for a fixed nominal pulse shape $h_p(t)$ (say, full-duty-cycle rectangular) and then accept any degradation which occurs if the actual $h_p(t)$ deviates from this nominal. It has been shown that for small values of σ_r/T such a receiver does not perform any worse than the equalizing type. When σ_r/T exceeds 0.25, the receiver sensitivity penalty can be a strong function of the specific receiver design. This will be illustrated again in Section 3.4.6.

3.4.5. Performance Calculations for Analog Receivers

In the digital receivers discussed above, performance was defined in terms of error probability. In many analog receiver applications, performance is defined in terms of the signal-to-noise ratio. Specifically, consider the analog receiver illustrated in Figure 3.37. The optical signal incident upon the photodetector is

$$P_r(t) = P_0[1 + k_m m(t)] \qquad \text{(watts)} \qquad (3.4.61)$$

where $m(t)$ is the message waveform having unity peak amplitude and zero average value, k_m is a modulation index having a value between zero and unity, and P_0 is then the average detected optical power, in watts.

In response to this signal, the detector emits a current which is a replica

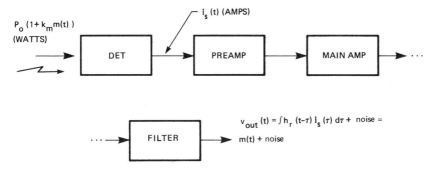

Figure 3.37. Analog receiver.

of $P_r(t)$ plus noise due to the statistics of photodetection and avalanche gain (if any). The preamplifier adds thermal noise to this signal.

The photodetector emits a displacement current made up of "bunches" of electron charges corresponding to the number of secondary electron–hole pairs produced by a given primary electron–hole pair. A single electron charge e produces an output response from the equalizer given by $eh_r(t - \tau)$, where e is the electron charge, $h_r(t)$ is the overall impulse response from the detector output to the equalizer output and τ is the generation time of the emitted electron. Thus the receiver–equalizer output is given by

$$v_{\text{out}}(t) = \sum_{-\infty}^{\infty} eM_k h_r(t - t_k) + n_{\text{th}}(t) \tag{3.4.62}$$

where M_k is the number of hole–electron pairs in the "bunch" of secondary pairs generated by primary pair k, t_k is the time at which primary k is generated, and n_{th} is preamplifier thermal noise of rms amplitude eZh_p, where h_p is the peak value of $h_r(t)$. From equation (3.4.62) one does not immediately see the explicit dependence of $v_{\text{out}}(t)$ upon the message $m(t)$. However, one must note that the average rate at which primary hole–electron pairs is generated is proportional to $P_r(t)$ given in (3.4.61). Using some standard mathematical tools for what are called compound Poisson processes[17] one can derive expressions for the mean and standard deviation of $v_{\text{out}}(t)$. One obtains the following:

$$\langle v_{\text{out}}(t) \rangle = eM \int \lambda(t') h_r(t - t') \, dt'$$

$$\sigma_v^2 = e^2 M^2 F_d \int \lambda(t') h_r^2(t - t') \, dt' + Z^2 e^2 h_p^2 \tag{3.4.63}$$

where

$$\lambda(t) = \frac{P_r(t)}{hf} = \frac{P_0}{hf}\left[1 + k_m m(t)\right]$$

and where M is the average avalanche gain, F_d is the avalanche detector excess noise factor, and σ_v is the standard deviation of $v_{\text{out}}(t)$.

If we average the mean squared noise, σ_v^2, over the probability distribution of $m(t)$ we obtain

$$\langle \sigma_v^2 \rangle = e^2 M^2 F_d \frac{P_0}{hf} \int h_r^2(t)\,dt + h_p^2 e^2 Z^2 \qquad (3.4.64)$$

If we now assume that $h_r(t)$ represents the impulse response of a low-pass filter having amplitude A and bandwidth B we obtain

$$\langle \sigma_v^2 \rangle = 2e^2 M^2 F_d \frac{P_0}{hf} A^2 B + (2AB)^2 Z^2 e^2 \qquad (3.4.65)$$

where we have used the fact that h_p is defined as the peak of the response $h_r(t)$, and equals $2AB$.

We assume that B is a bandwidth adequate to pass the modulation $m(t)$. Ignoring a dc term we obtain from (3.4.63)

$$\langle v_{\text{out}}(t) \rangle = \frac{eMP_0 A k_m m(t)}{hf} \qquad (3.4.66)$$

We can now calculate the ratio of peak signal $[m(t) = 1]$ squared to mean squared noise, a signal-to-noise ratio (SNR):

$$\text{SNR} = \frac{\langle v_{\text{out}}(t) \rangle^2|_{m(t)=1}}{\langle \sigma_v^2 \rangle} = \frac{k_m^2 P_0^2/(hf)^2}{(2P_0 F_d B/hf) + (4Z^2 B^2/M^2)} \qquad (3.4.67)$$

Consider first some special cases of (3.4.67). If preamplifier thermal noise is negligible and we, therefore, set $Z = 0$, $M = 1$, and $F_d = 1$, we obtain the analog quantum limit

$$\text{SNR}_{\text{quantum limit}} = \frac{k_m^2 P_0}{2hf B} \qquad (3.4.68)$$

If we assume that the second term in the denominator of (3.4.67) dominates, then we have the thermal noise limit

$$\text{SNR}_{\text{thermal noise limit}} = \frac{k_m^2 P_0^2 M^2}{4(hf)^2 B^2 Z^2} \qquad (3.4.69)$$

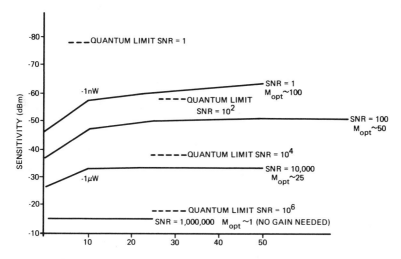

Figure 3.38. Receiver sensitivity vs. avalanche gain for various values of SNR optical fiber.

As one increases the avalanche gain M, one reaches a point where the first term in the denominator of (3.4.67) dominates.

Since F_d is in general an increasing function of M, one finds that there is an optimal gain that maximizes the signal-to-noise ratio. The higher the desired signal-to-noise ratio, the lower the optimal gain. Figure 3.38 shows curves of the quantity P_0 in dBm as a function of avalanche gain M for various values of SNR, for a typical 25-MHz bandwidth receiver with $Z = 1400$. The quantum limits are also shown. The detector ionization ratio parameter k is assumed to be 0.025, and the modulation index k_m is assumed to be 0.7.

In equation (3.4.61) we have assumed linear intensity modulation. That is, the optical power is proportional to the message $m(t)$. Other analog modulation formats such as subcarrier FM will be described in Chapter 4. Also sources of noise related to the transmitter have been ignored here, but will be considered in Chapter 4.

3.4.6. Example of a Practical Receiver

Figure 3.39 shows a practical 50 Mb s^{-1} digital receiver used in a fiber optical field experiment.[18] The preamplifier is shown in detail in Figure 3.40. It consists of a grounded-emitter transistor stage followed by an emitter follower impedance converter. The preamplifier is a trans-impedance type with a thermal noise dominated by the Johnson noise of the feedback resistor and the shot noise of the base bias current of Q_1.

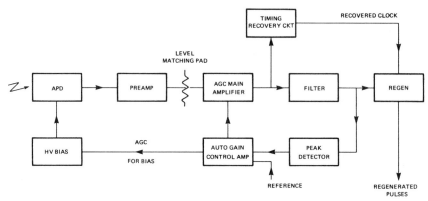

Figure 3.39. Practical 50-Mb/s digital receiver.

The open-loop gain of the amplifier from the input of the first transistor to the output of the second transistor, A_{forward}, is

$$A_{\text{forward}} = 4000/R_{e1} \tag{3.4.70}$$

where R_{e1} = emitter resistance of $Q_1 = kT/eI_{e1}$, I_{e1} is the average emitter current of Q_1, $kT = 4 \times 10^{-21}$, $e = 1.6 \times 10^{-19}$.

From the biasing shown we have (assuming a base-emitter drop of 0.7 V/transistor)

$$I_{e1} \cong (5 - 1.4)/4000 = 0.9 \text{ mA} \tag{3.4.71}$$

Thus

$$A_{\text{forward}} \approx 3.6e/kT \approx 144 \tag{3.4.72}$$

The input capacitance of the first transistor plus the detector capacitance is about 5 pF. Thus the feedback ratio at 50 MHz is

$$\beta = \frac{Z_{\text{in}}}{Z_{\text{in}} + Z_f} \approx \frac{Z_{\text{in}}}{Z_f} = [2\pi \times 50 \times 10^6 \times 5 \times 10^{-12} \times 4000]^{-1}$$

$$= \frac{1}{2\pi} \cong 0.16 \tag{3.4.73}$$

Thus the loop gain at 50 MHz is about 23.

The noise at the input from the feedback resistor is $4kT/4000 \text{ A}^2/\text{Hz}^{-1}$. The noise at the input from the dc base current of Q_1 is $2eI_{b1} \text{ A}^2/\text{Hz}$, where I_{b1} is the base current. If we assume a current gain of about 300 in Q_1,

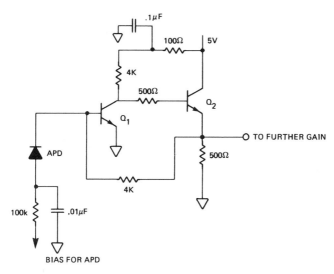

Figure 3.40. Practical preamplifier (detail).

then from (3.4.71) we obtain I_{b1} approximately equal to 3 μA of dc bias, and we have

$$4kT/4000 \approx 4 \times 10^{-24}(\text{A}^2/\text{Hz}) \quad \text{(noise from } R_f)$$

$$2eI_{b1} \approx 1 \times 10^{-24}(\text{A}^2/\text{Hz}) \quad \text{(base bias shot noise)} \quad (3.4.74)$$

The equivalent physical input noise resistance resulting from the sum of these noise sources is about 3200 Ω. From equation (3.4.7a) we have (setting $B = $ the half baud $= 25$ MHz)

$$Z = \left(\frac{4kTB}{3200} \right)^{1/2} \bigg/ eB = 3000 \quad (3.4.75)$$

The transimpedance preamplifier shown can produce a peak-to-peak output signal of about 1 V, limiting the maximum current input to 250 μA, if distortion is to be avoided.

The preamplifier is followed by an automatic gain control (AGC) amplifier with a maximum input signal level of 200 mV. In order to maximize the dynamic range of the overall receiver, a voltage divider is placed between the preamplifier output and the AGC amplifier input, matching the overload levels. This results in some noise penalty (about 1 dB in receiver sensitivity in this case) but a 7-dB increase in allowed optical signal level before overload. Thus the optical dynamic range is increased by about

6 dB. The avalanche photodiode gain is part of the AGC loop. As the optical signal level is increased from its minimum value, the APD bias voltage is first gradually reduced to a minimum allowable level which guarantees APD sweepout (fast response speed). The AGC amplifiers are then reduced in gain, until the input to the first AGC amplifier overloads. Beyond this, further signal increases cause distortion.

With the value of Z given above, and with the APD used ($k = 0.04$), the minimum number of detected photons per pulse for an a 10^{-9} error rate is about 500, at an optimal avalanche gain of about 100.

The average detected optical power with half marks and half spaces is then (at maximum sensitivity)

$$P_{av} = 500hf \times 50 \times 10^6/2 = 2.5 \times 10^{-9} \, \text{W} \cong -56 \, \text{dBm}$$

The detector output current at maximum sensitivity is

$$i_{out} = 2.5 \times 10^{-9} \, \text{W} \times e/hf \times 100 \, (\text{avalanche gain}) =$$
$$= 2 \times 10^{-7} \text{A}$$

The preamplifier output voltage at maximum sensitivity is therefore about 0.8 mV.

The avalanche photodiode is guaranteed to have a gain of less than 10 at the minimum allowed voltage for sweepout. Thus assuming the gain is in fact 10 when the back-bias voltage is reduced (compared to 100 at maximum sensitivity) we see that we can increase the optical signal level about 40 dB from the minimum value before the preamplifier output voltage reaches the overload value of 1 V. Thus the receiver dynamic range is about 40 dB.

In actual operation, one must allow some margin at the lower optical power end to guarantee the required error rate (10^{-9}) in spite of component variations with temperature and time. Also margin must be allowed for power line noises, improper alignment and other contingencies. Thus the actual sensitivity with margin is about -50 dBm at an optimal APD gain of 50, and the actual dynamic range is about 32 dB.

It is interesting to write down a typical specification for a receiver for a 50-Mb/s system

Receiver Specification (Specimen)

1. All specifications below will be valid over the temperature range -20–$60°$C.

2. The required power supply voltages will be $+5$ V and -5.2 V $\pm 5\%$ and high voltage for the APD as discussed below. The allowed power line noise will exceed 50 mV rms distributed arbitrarily in the frequency band 0–50 kHz.

3. The receiver will incorporate an avalanche photodiode detector. The detector quantum

efficiency (excluding coupling loss to be described below) will exceed 90% at 830 nm wavelength. The detector ionization ratio parameter will not exceed 0.04. The detector will be capable of gains of up to 120 without microplasma breakdown effects. In particular the detector excess noise factor will not exceed the theoretical excess noise (for a detector with an ionization ratio of 0.04) by more than 10% with gains of up to at least 120. The APD bias voltage required for a gain of 120 will not exceed 450 V over the entire temperature range. A voltage v_{min} must be specified, independent of temperature, and the same for all detectors such that for an APD bias equal to v_{min} the detector is "swept out" and has a gain of less than 10.

The detector dark current will not exceed 10^{-9} A at a gain of 100. The detector 3-dB bandwidth will exceed 150 MHz at all gain values.

4. The receiver will incorporate a fiber pigtail permanently attached to the detector. The fiber will be a graded index type with a numerical aperture (N.A.) exceeding 0.2, a core diameter exceeding 62.5 μm, and a cladding diameter of 125 $\pm$ 6 μm. The cladding axis will be centered on the core axis with an offset of no more than 3 μm. The coupling loss at the interface of the pigtail and the detector (including reflections) will not exceed 1 dB, assuming all fiber modes are uniformly excited at a wavelength of 830 nm.

5. The receiver will incorporate a preamplifier having a transimpedance of 5000 Ω. The low-frequency cutoff (if ac coupled) of the preamplifier will not exceed 5000 Hz. The high-frequency 3-dB rolloff will be above 75 MHz. Within the passband 50 kHz to 25 MHz, the frequency response will be flat to within ± 0.25 dB. When bandlimited with a "full raised cosine" roll-off to 50 MHz (6 dB down at 25 MHz), the amplifier will have a noise parameter Z of less than 3000 (at 20°C). (This corresponds to an equivalent flat current noise source at the input equal to that of a 3000-Ω resistor.)

6. The receiver will not overload (2% compression) for output signals of less than 1 V. The output impedance will be less than 50 Ω.

The above specification was presented to illustrate the typical parameters which must be specified to define a receiver. An actual specification would contain more detail as to how these parameters are to be measured, and might, of course, specify different parameter values.

A final comment on the receiver shown in Figure 3.39 concerns the effect of pulse spreading, due to delay distortion in the fiber, on the receiver sensitivity. Since the receiver incorporates an automatic gain control, if the optical input level is increased from the minimum allowed level, the avalanche gain is reduced by the feedback control circuitry. If the incoming optical pulse stream suffers from intersymbol interference, the level of the optical signal must be increased, to compensate for the corresponding loss of noise margin at the receiver output. This increase in optical signal level causes the avalanche gain to drop (because of the AGC). The drop in avalanche gain in this situation is undesirable, tending to increase the penalty in receiver sensitivity due to intersymbol interference, relative to the penalty associated with an equalizing receiver which automatically reshapes the output pulses and optimally adjusts the avalanche gain. Thus although all receivers must suffer some penalty from input intersymbol interference, this receiver tends to perform worse than theory requires. (For the application contemplated intersymbol interference due to delay distortion in the fiber was not expected to be a problem.)

In order to sample the receiver output to produce a "regenerated digital pulse stream, one requires a periodic clock signal which is synchronized with the time slots of the receiver output pulses. Such a clock signal can be derived from the receiver itself using a phase-locked loop timing recovery circuit. The receiver provides an output from the main AGC amplifier as shown in Figure 3.39. This signal is differentiated and rectified to produce a new signal which is well known to contain a strong discrete component at the baud. This component is filtered from the rest of the spectrum with a narrow-bandwidth phase-locked loop. There is a tradeoff between the amount of filtering which can be allowed in the PLL and the ability of the PLL to capture (lock on to) the desired component and to track phase modulation on the desired component.[19] Variations in the pattern of pulses incident on the receiver result in phase jitter of the derived clock signal relative to the positions of the incoming time slots. This jitter will accumulate when many repeaters are placed in tandem. The origin of the jitter is mainly from imperfections in the phase-locked loop circuits (dc offsets) and interference between adjacent pulses in the main amplifier output signal. The subject of timing recovery and jitter accumulation has been treated thoroughly in the references.[19-21]

3.5. Fiber Subsystems [4]

In this section we shall study the fiber as a subsystem in terms of its input/output relationships. We shall model the fiber as an optical power filter and define the important parameters which characterize the fiber. Then we shall discuss measurements of these parameters.

3.5.1. Fiber Input/Output Properties

The fiber accepts, as its input, power from an optical source. In response to this, power is emitted at the fiber output. A key assumption used in modeling optical fiber input/output relationships is "base-band linearity."[22] That is

$$p_{out}(t) = \int h_{fiber}(t - u) \, p_{in}(u) \, du \qquad (3.5.1)$$

where $p_{out}(t)$ is the fiber output power (watts), $p_{in}(t)$ is the fiber input power (watts), and $h_{fiber}(t)$ is the fiber base-band impulse response.

Equation (3.5.1) seems innocuous enough. It states that as a result of delay differences associated with different paths through the fiber (and different group velocities of different wavelengths within the source spectrum) the output power is a superposition of weighted and delayed versions of the input power.

From the linearity of Maxwell's equations it follows that the output field must be a linearly filtered version of the input field. However, since the power is proportional to the square of the field, it is not obvious that (3.5.1) is valid in general. Indeed, there are situations where a system designed under assumption (3.5.1) might not work as anticipated. This is the key. It is not relevant to system design whether (3.5.1) is exactly valid. What matters is whether (3.5.1) is a good enough approximation so that the system will work as predicted. Thus to analyze the validity of (3.5.1) one must be careful to ask the right questions. The subject of base-band linearity is too mathematically complex to treat here in general.[22] However, before stating some general results of studies in the references, it is useful to give one simple (but nontrivial) example of when (3.5.1) is clearly valid.

Consider a multimode waveguide with k orthogonal modes (by definition). Assume that as a field propagates along the waveguide, no exchange of energy occurs between the modes. Although the group velocity is different for different modes, assume that for a given mode there is no delay distortion (pulse spreading due to dispersion). Let the two-dimensional parameter, s, represent two-dimensional position in the fiber cross section. Let the complex amplitude of mode j at the fiber input be proportional to the square root of a (positive) message $m(t)$. Thus we have the real input field, $E_{in}(s, t)$ given by

$$E_{in}(s, t) = 2^{1/2} \operatorname{Re}\{\varepsilon_{in}e^{i2\pi ft}\} = 2^{1/2} \operatorname{Re}\left\{[m(t)]^{1/2} e^{i2\pi ft} \sum_{1}^{N} a_k\psi_k(s)\right\} \quad (3.5.2)$$

where $\psi_k(s)$ is the spatial mode k as a function of position, s; f is the optical frequency; a_k is the complex amplitude of mode k; $\operatorname{Re}\{\ \}$ is the real part of $\{\ \}$; and $\varepsilon_{in}(s, t)$ is the complex input field.

As each mode propagates it is delayed by τ_k. Thus the real output field is given by

$$E_{out}(s, t) = 2^{1/2} \operatorname{Re}\{\varepsilon_{out}(s, t) e^{i2\pi ft}\}$$

$$= 2^{1/2} \operatorname{Re}\left\{e^{i2\pi ft} \sum_{k=1}^{N} a_k\psi_k(s) [m(t - \tau_k)]^{1/2}\right\} \quad (3.5.3)$$

The power at the input or output is obtained by taking the square of the complex field magnitude and integrating the result over the fiber cross section. That is,

$$p_{in}(t) = \int \varepsilon_{in}(\rho, t) \varepsilon_{in}^*(\rho, t) d^2p = m(t) \sum_{k=1}^{N} \sum_{j=1}^{N} \int a_k a_j^* \psi_k(s) \psi_j^*(s) d^2s \quad (3.5.4)$$

but since the modes are orthonormal, i.e.,

$$\int \psi_k(s)\,\psi_j^*(s)\,ds = 1 \qquad \text{for } k = j$$

$$= 0 \qquad \text{for } k \neq j \qquad (3.5.5)$$

we obtain

$$p_{\text{in}}(t) = m(t) \sum_{k=1}^{N} |a_k|^2 \qquad (3.5.6)$$

Similarly, at the output we obtain

$$p_{\text{out}}(t) = \sum_{1}^{N} m(t - \tau_k) |a_k|^2 = \sum_{1}^{N} \gamma_k p_{\text{in}}(t - \tau_k) \qquad (3.5.7)$$

where

$$\gamma_k = |a_k|^2 \left/ \sum_{j=1}^{N} |a_j|^2 \right.$$

Comparing (3.5.7) to (3.5.1) we see that in this special case the output power is indeed a weighted superposition of delayed versions of the input power. That is, base-band linearity holds.

What was key in the above analysis was the dropout of certain "cross terms" in the square of the output complex field, because of the orthogonality of the various modes. In all analyses of base-band linearity there are analogous cross terms which must be shown to be negligible in their effect on the system. As a general rule, base-band linearity is a good approximation for systems where the optical source bandwidth is very large compared to the base-band bandwidth of the receiver. Under such circumstances, the phase variations of the source, with time, tend to average out the undesired cross terms. Thus where LED sources or injection lasers with relatively low coherence are used, base-band linearity is generally valid. With coherent laser sources one can construct counterexamples where base-band linearity is a bad approximation. However, the significance in practical applications of such counterexamples is doubtful. For the remainder of this book we shall assume (3.5.1) to be valid.

Figure 3.41 shows some typical examples of impulse responses $h_{\text{fiber}}(t)$ which one might expect to observe in actual fibers. Response (a) is approximately rectangular in shape, and is what one might expect to see for a step index fiber with equal attenuation for all modes and no mode coupling. Response (b) might correspond to a step index fiber with no mode coupling but with an increasing attenuation for higher-order modes. Response (c) might correspond to a graded index fiber with no mode coupling but with

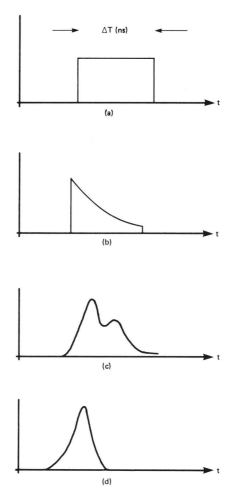

Figure 3.41. Typical impulse responses.

an imperfect grading. The residual impulse response shows the effects of random imperfections in the index-grading process. Response (d) might correspond to a fiber with significant mode coupling along its length. It is interesting to discuss the reasons for the above pulse shapes.

Pulse shape (a) is representative of a fiber with many (N) modes having propagation delays which are roughly uniformly spread over a range $\Delta\tau$ (ns km^{-1}). If the modes are equally excited at the fiber input, then a narrow pulse at the input will produce a comb of output pulses as shown in Figure 3.42. If we assume that the detector has finite bandwidth, then this comb of output pulses will smooth out to the impulse response shown

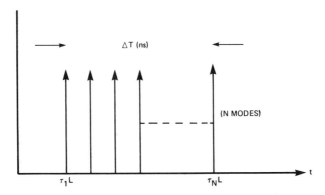

Figure 3.42. Impulse response (detail).

in Figure 3.41a. If the delay of the fastest mode is τ_1 (ns km^{-1}) and the delay of the slowest mode is τ_N then the width of the impulse response is simply given by

$$W = (\tau_1 - \tau_N) L = \Delta\tau L \qquad \text{(nanoseconds)} \qquad (3.5.8)$$

where L is the fiber length in kilometers.

The corresponding rms width of the impulse response of Figure 3.41a defined in (3.4.60) is

$$\sigma_r = \frac{\Delta\tau L}{(12)^{1/2}} \qquad (3.5.9)$$

If higher-order (angle, for step index fibers) modes have more loss, as is often the case, then the response of a multimode fiber to a narrow input pulse, which excites all modes uniformly, could look as shown in Figure 3.43, where for simplicity we assume that modes with more delay have proportionately more loss. Thus the band-limited impulse response (smoothed) looks like 3.41b. If the fastest mode has delay τ_1 (ns km^{-1}) and attenuation α_1 (Np km^{-1}) and if incremental attenuation is proportional to incremental delay, then we have (for uniform excitation of all modes)

$$h_{\text{fiber}}(t) = C_0 \sum_{k=1}^{N} \delta(t - \tau_k L) \, e^{-\alpha_k L}$$

$$= C_0 \sum_{k=0}^{N-1} \delta(t - \tau_1 L - k\,\Delta\tau L) \, e^{-(\alpha_1 L + k\,\Delta\alpha L)} \qquad (3.5.10)$$

where C_0 is a normalizing constant, $\delta(t)$ is the Dirac "delta function," $\Delta\tau$ is

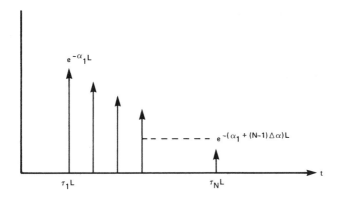

Figure 3.43. Impulse response (detail).

the incremental delay between modes, and $\Delta\alpha$ is the incremental attenuation between modes. Smoothing out the impulses one obtains

$$h_{\text{fiber}}(t) = \frac{C_0}{\Delta\tau} e^{-\alpha_1 L} e^{-\Delta\alpha(t-\tau_1 L)/\Delta\tau} \qquad \text{for } \tau_1 L \le t \le \tau_1 L + (N-1)\Delta\tau L$$

$$(3.5.11)$$

For a long fiber with $N\Delta\alpha L \gg 1$ we obtain the corresponding rms width

$$\sigma_r = \Delta\tau/\Delta\alpha \qquad \text{(nanoseconds)} \qquad (3.5.12)$$

Thus with differential attenuation of higher-order modes directly proportional to the differential delay, the rms pulse width converges to a constant independent of the length of the fiber. In a real fiber, this direct proportionality of differential delay and differential attenuation may not be valid, but the effect, i.e., a less than linear increase of pulse rms width with length, might still apply if modes with longer delay have more attenuation.

For a graded index fiber, without mode coupling, ideally the impulse response would look like Figure 3.41a; but with a much narrower pulse width than a step index fiber with the same N.A. (numerical aperture). However, because of unavoidable errors in the index grading, certain groups of guided modes may exist, which have distinctly different propagation delays than other groups. It is possible for the propagation delays of these distinct groups to be separated by gaps where no modes of a given delay exist. Thus a narrow input pulse exciting, say, two groups of such modes, can produce a response as shown in Figure 3.41c. The amplitude of the two peaks relative to each other would depend upon the excitation conditions. Fibers have been observed to have impulse responses consisting of several distinct peaks.

Another interesting phenomenon in fiber input/output characterization is mode mixing.[23] To understand mode mixing, consider a two-mode fiber where mode 1 has delay τ_1 (ns km^{-1}) and mode 2 has delay τ_2 (ns km^{-1}). Assume that these modes are equally excited and that the fiber length is L (km). The resulting impulse response would be as shown in Figure 3.44.

The pulse shown has an rms width proportional to the fiber length and given by

$$\sigma_r = \frac{(\tau_2 - \tau_2)}{2} L = \frac{\Delta\tau L}{2} \tag{3.5.13}$$

Now suppose the fiber is cut in half and a "mode mixer" is inserted. The definition of the mode mixer is that energy propagating in mode 1 is split equally between modes 1 and 2, and vice versa. The resulting impulse response is shown in Figure 3.45. The corresponding rms width is given by

$$\sigma_r = (\Delta\tau L/2)/2^{1/2} \tag{3.5.14}$$

If instead of one mode-mixing point, suppose there are $(N - 1)$ equally spaced mode mixing points, where N is the ratio of the fiber length to the spacing between mode mixers. Then the corresponding rms width is

$$\sigma_r = (\Delta\tau L/2)/N^{1/2} \tag{3.5.15}$$

Thus the rms width is reduced by $N^{1/2}$. It can be shown, in general, that in a multimode fiber with mode coupling uniformly distributed along its length, the rms width of its impulse response is reduced, from the case of no mode coupling, by a factor $N^{1/2}$, where

$$N = L/L_c \tag{3.5.16}$$

Here L is the fiber length and L_c is the average distance of propagation required for energy to randomize amongst the modes ("coupling length").

Furthermore, if L is much greater than L_c the fiber impulse response

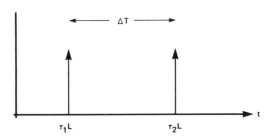

Figure 3.44. Two-mode fiber response.

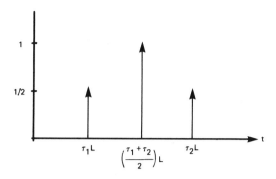

Figure 3.45. Two-mode fiber response with a mode mixer in the middle.

tends to converge to a Gaussian shape as shown in Figure 3.41d, i.e., of the following form:

$$h_{\text{fiber}}(t) = \frac{1}{(2\pi)^{1/2}\,\sigma_r}\,e^{-t^2/2\sigma_r^2} \tag{3.5.17}$$

where

$$\sigma_r = \sigma_0 L_c^{1/2} L^{1/2}$$

σ_0 is the rms pulse spreading per unit length without mode mixing and L_c is the coupling length.

Thus with uniform mode mixing, the pulse width increases as the square root of the fiber length (rather than linearly with length), for lengths longer than L_c (the coupling length). Figure 3.46 shows a curve (hypothetical) of the rms width vs. length for a fiber with mode mixing, which can be obtained by successively cutting the fiber back. (Fiber measurements will be discussed in Section 3.5.2.)

From (3.5.17) and the discussion above, it would seem desirable to increase mode coupling to reduce the rms width of the fiber impulse response. Such mode coupling can be caused by random or intentional fluctuations in the fiber geometry with position along its length (e.g., bending, diameter changes in the core, etc.) at mechanical periods of around 1–10 mm.

Unfortunately the same mechanisms which cause guided modes to couple to each other also cause guided modes to couple to the radiation field, introducing increased loss. Thus there is an added loss proportional to the ratio of the fiber length L to the coupling length L_c, which can be

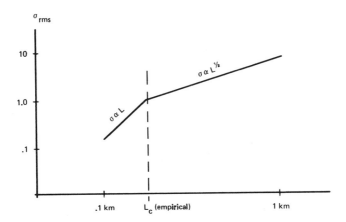

Figure 3.46. Fiber rms width vs. length with mode mixing.

expressed in nepers per coupling length. That is there is an excess loss α_c given by

$$\alpha_c = \alpha_{\text{const}} L/L_c \qquad \text{(nepers)} \qquad (3.5.18)$$

where α_{const} is a constant which depends upon the type of fiber (graded, step) and the spectrum of the mechanical perturbations causing the mode mixing. All the former things being equal, so that α_{const} is a constant, the more mode mixing (larger amplitude perturbations), and therefore the shorter L_c, the more loss. Thus as we shall see in Chapter 4, in a system, there is a tradeoff between increasing mode mixing to reduce pulse spreading in transmission through the fiber, and excess loss due to the mode mixing.

As a complement to the impulse response, fibers are often characterized by a power transfer function $H_{\text{fiber}}(f)$ which is the Fourier transform of the impulse response. If base-band linearity holds [equation (3.5.1)] then one should be able to measure either one directly (as will be described in Section 3.5.2) and to obtain the other by transforming the data:

$$H_{\text{fiber}}(f) = \int h_{\text{fiber}}(t) \, e^{-i2\pi ft} \, dt \qquad (3.5.19)$$

It is interesting to see the relationship between certain parameters in the impulse response and the transfer function. Consider the rms width, σ_r.

It follows from (3.5.19) that the rms width is the coefficient of a parabolic approximation of the transfer function. That is

$$H_{\text{fiber}}(f) = \int h_{\text{fiber}}(t)\, e^{-i2\pi ft}\, dt \tag{3.5.20}$$

$$= \int h_{\text{fiber}}(t) \left[1 - i2\pi ft + \frac{(i2\pi ft)^2}{2!} \cdots \right] dt$$

$$\cong \int h_{\text{fiber}}(t) \left[1 - i2\pi ft - \frac{(2\pi ft)^2}{2} \right] dt$$

$$= H_{\text{fiber}}(0) \left[1 - i2\pi f\bar{t} - (2\pi f)^2\, \overline{t^2}/2 \right]$$

where

$$H_{\text{fiber}}(0) = \int h_{\text{fiber}}(t)\, dt \quad \text{and} \quad \overline{t^n} = \int h_{\text{fiber}}(t)\, t^n\, dt \big/ H_{\text{fiber}}(0)$$

Therefore

$$\left| H_{\text{fiber}}(f) \right| \cong H_{\text{fiber}}(0) \left[1 - (2\pi f\sigma_r)^2/2 \right]$$

where

$$\sigma_r^2 = \overline{t^2} - (\bar{t})^2$$

Consider a fiber impulse response as shown in Figure 3.47. The "tail" could correspond, for example, to energy which temporarily couples to slower cladding modes and then back again to core modes, while propagating along the fiber. Such tails are not unusual in actual fiber impulse responses. In the corresponding transfer function shown in Figure 3.48, the

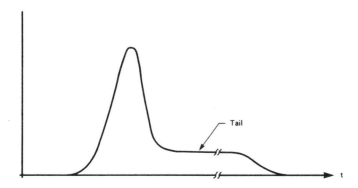

Figure 3.47. Fiber impulse response with a tail (exaggerated).

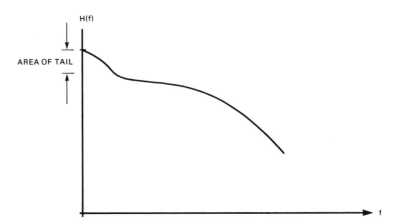

Figure 3.48. Transfer function corresponding to Figure 3.47.

tail shows up as a low-frequency bump. The impact of such tails on system performance will be discussed in Chapter 4.

Figure 3.49 shows the transfer function corresponding to an impulse response like the one shown in Figure 3.41c. The frequency of the notch in the transfer function is proportional to the inverse of the spacing in time between the peaks in the impulse response.

It is interesting to consider the overall impulse response and the transfer function of a fiber formed by two fibers placed in tandem.

If there is no mode coupling along the fibers individually, and if the splice couples modes in fiber 1 to the corresponding modes in fiber 2, then the impulse response of the overall fiber should be as wide as the sum of

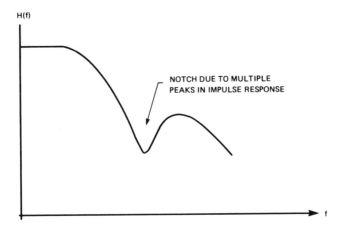

Figure 3.49. Transfer function corresponding to Figure 3.41c.

the widths of the impulse responses of the pieces. If, on the other hand, the fibers have mode mixing along their lengths and/or mode mixing occurs at the splice point, then the overall impulse response is obtained by convolving the individual impulse responses:

$$h_{\text{fiber overall}}(t) = \int h_{\text{fiber 1}}(u)\, h_{\text{fiber 2}}(t-u)\, du \qquad (3.5.21)$$

and (without any approximations)

$$\sigma_{\text{overall}\, r}^2 = \sigma_{1r}^2 + \sigma_{2r}^2 \qquad (3.5.22)$$

Thus with mode mixing, the overall impulse response has an rms width equal to the square root of the sum of the squares of the rms widths of the pieces (regardless of the shapes of the impulse responses of the pieces). The overall transfer function in this case is the product of the individual transfer functions, as (3.5.22) would predict.

There has been some experimentation with a delay compensation approach in which two imperfect graded index fibers are spliced together in an effort to obtain a total fiber which is better in its delay distortion characteristics then one might expect from the individual pieces. For example, if, because of grading errors, a certain mode group in one fiber has a longer than ideal delay per unit length, and if in the other fiber the corresponding mode group has a shorter than ideal delay per unit length, then by splicing the fibers (without mode mixing) these delay deviations can partially cancel. Whether such compensation is practical or not depends upon the application. It would seem that in the long run a better approach would be to make fibers with closer to ideal index gradings. However, in special applications, where delay distortion is critical, one could resort to this mix-and-match technique. Another mix-and-match application involves loss. In a tandem link made up of cables with fibers of varying losses (due to normal manufacturing deviations) one could try to match high-loss fibers in one section to low-loss fibers in another, in order to obtain a more uniform loss amongst the fibers in the overall spliced cable.

A major problem in characterizing optical fibers is associated with the multimode nature of these waveguides. Typical graded index fibers have several hundred modes with different losses, and different delays per unit length. The impulse response shape and the fiber loss is therefore a function of the launch conditions at the fiber input and the mode mixing along the fiber cable. A fiber with a lot of mode mixing tends to be less sensitive to the launch conditions, because the energy is randomized amongst the modes in a distance which may be short compared to the overall fiber length. However, since mode mixing is associated with loss, efforts to make lower loss cables has led to less and less mode mixing. Correspondingly, the input/output properties of such fiber cables is increasingly dependent upon the type of optical source exciting the fiber and the launch conditons. This will be treated further below.

3.5.2. Measurements of Fiber Parameters†

In addition to mechanical parameters, which will not be discussed in this book, the two major input/output parameters characterizing a fiber are the loss and the impulse response. Measurements of these parameters will be the subject of this section.

3.5.2.1. Loss Measurements. The measurement of loss is complicated by the fact that, as mentioned above, typical fibers have hundreds of modes with varying loss. Indeed, in the research laboratory, where one is trying to produce lower loss fibers, there is a temptation when measuring loss to use coherent sources and to launch only those modes which tend to have the lowest losses. A complete characterization of the fiber would give measured loss as a function of which modes are excited at the input. However, such a characterization would be difficult to obtain. A more practical approach is to excite all modes equally using an incoherent source or using a mode scrambler at the launch end. A mode scrambler is any device which scatters light from a coherent source randomly amongst the modes at the fiber input. It has been found that measurements made in different places with mode scramblers at the launch end lead to very consistent results.

There are several methods used to make loss measurements. One method is shown in Figure 3.50. In this approach an optical transmitter and an optical receiver are first interconnected with a short "strap" fiber, assumed to have negligible loss. The receiver gain is adjusted to "zero" a meter, in this case using an adjustable transimpedance. The strap is then

---

† See Chapter 11 of Reference 4 and References 24–26.

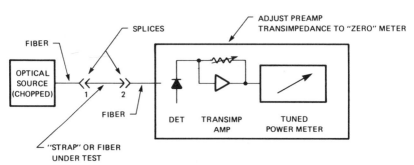

Figure 3.50. Loss measuring set.

removed, and replaced by the "fiber under test." The receiver transimpedance is increased to rezero the meter. The fiber loss is given by the ratio of the transimpedance at the receiver before and after replacement of the strap with the fiber under test. [Fiber loss in decibels is 10 log (transimpedance ratio).] One assumption always being made in a loss measurement like this one is that the losses coupling into the "strap" at splice 1, and out of the "strap" at splice 2, are the same as the corresponding losses with the fiber under test. To reduce the uncertainty in this assumption, sometimes the measurement is made as follows. The transimpedance is adjusted to "zero" the meter with the fiber under test in place. Then the fiber under test is disconnected at splice 2 only, and a small piece of the fiber under test is broken off near splice 1. This creates a "strap" fiber without disturbing splice 1. Thus the splice at the launch end should have the same loss in both the "strap" and fiber-under-test measurements. The optical source used in the above measurement can be an LED or a laser. For accurate measurements a PIN detector is generally used. Portable instruments of this type have been built with dynamic ranges of 0–50 dB+ of fiber loss and precisions of ± 0.1 dB. Accuracy of the loss measurement is controlled by the splices or connectors (in a well-designed instrument) and is typically better than ± 1 dB.

The above "substitution method of loss measurement" does not give any detail as to how uniform the fiber loss is along its length. An alternative measurement which gives such detail utilizes the "optical time domain reflectometer."[25,26] This device is shown schematically in Figure 3.51. A transmitter produces pulses of light periodically, along with a trigger signal. These pulses are launched into a fiber using a directional coupler. Light reflected by the fiber is directed to an optical receiver with a bandwidth compatible with the width of the transmitted pulses. The receiver output is

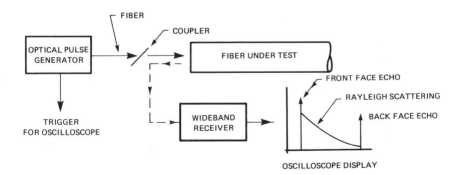

Figure 3.51. Optical time domain reflectometer (OTDR).

displayed on an oscilloscope triggered by the transmitter. One might expect to see only two reflections, one from the launch end, and one from the output end of the fiber. However, with a sensitive receiver one sees a trace like the one shown in Figure 3.52. The discrete reflections are the ones described above. The continuous, decaying part of the trace is Rayleigh scattering which has been recaptured by the fiber in the backward direction. The decay of this scattering represents the round trip attenuation to a given point along the fiber (where scattering occurs) and back. Thus if the level of Rayleigh scattering is roughly uniform along the fiber, the decay of the continuous part of the echo trace represents the loss vs. length of the fiber. Figure 3.53 shows the "echo scan" of a composite fiber made up of two fiber pieces spliced together. At the splice point there is a discrete reflection. In addition, the step down in the level of the continuous Rayleigh scattering trace at the splice point represents the round trip loss through the splice. Analyses of the details of signals and noise in optical time domain reflectometers (OTDRs) can be found in the references. The fundamental limitations on the range of loss one can "look through" with an OTDR are noise, source output pulse fluctuations, and "pickup."

The noise is caused by the low level of returned Rayleigh scattering which, as mentioned, is additionally attenuated by the loss in returning from the scattering point back to the receiver. To enhance the measurement sensitivity one can use a "boxcar averager." Such a device samples and averages the echos from many transmitted pulses. A difficulty in the averaging process is the variation in the energy of the transmitted pulses from pulse to pulse. To remove this variation, one can normalize the averages of samples of the echo vs. time to the averages of samples taken at a fixed time along the echo trace. That is, each echo is sampled twice, once at a fixed reference point and then again at a varying time point. To remove pickup, one can subtract off the boxcar-averaged receiver output which

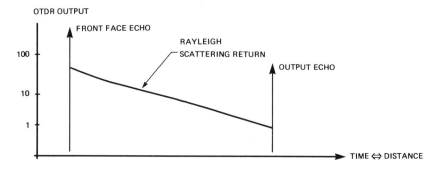

Figure 3.52. OTDR output detail.

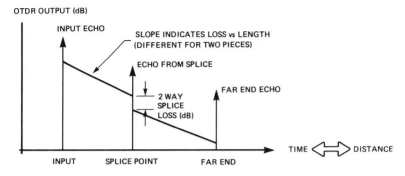

Figure 3.53. Echo scan of composite fiber.

results when the optical signal is removed from the fiber. Using these techniques, one can accurately, reproducibly, and nondestructively measure fiber loss vs. position, and splice losses, through more than 25 dB of one-way attenuation, with a precision of better than 0.1 dB.

One can also use the OTDR to locate breaks in fibers. Using very narrow pulses (less than 500 ps FWHM) obtained from suitably designed laser transmitters (see Section 3.3.3 above and Figure 3.11) one can resolve echos in short fibers with a precision of less than 10 cm. For example, if one suspects that a defective transmitter or receiver module, with an optical pigtail, has a damaged pigtail fiber, and if the fiber is armored by an opaque cable, then one can use the OTDR to determine whether the module can be repaired by locating the break.

3.5.2.2. Impulse Response Measurements.† To measure impulse response one can either work directly in the time domain or in the frequency domain (Chapter 4 of Reference 4). Figure 3.54 shows a time domain impulse response measurement set. The transmitter produces narrow pulses (0.5 ns) at a repetition rate of about 10 kHz (see Section 3.3.3 above and Figure 3.11). It also produces a trigger signal for the receiver. The transmitter output pulses are launched into a fiber under test or a "strap" fiber. The trigger delay generator delays the trigger by an amount equal to the propagation delay through the fiber (5 μs km^{-1}). The fiber output is detected, and displayed on a sampling oscilloscope synchronized by the delayed trigger pulses. The oscilloscope "sample output" signal can be converted to digital form and averaged further using a minicomputer. The resulting averaged pulse response and its Fourier transform can be displayed or printed out. The

† See Chapter 11 of Reference 4.

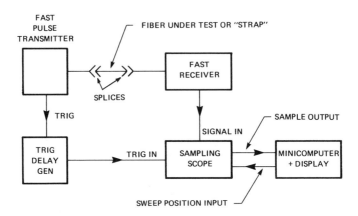

Figure 3.54. Impulse response measuring set.

transfer function of the fiber under test is obtained by dividing the Fourier transform of the averaged pulse response obtained using the "fiber under test" by the Fourier transform of the averaged pulse response obtained with the strap fiber.

The main limitations on the accuracy and the reproducibility of these measurements are (in addition to the dependence on launch conditions) detector nonlinearity, jitter in the delayed trigger, noise, pickup, laser pulse-to-pulse variations, and laser spectral width.

When one normalizes the Fourier transform obtained with the fiber under test to that obtained with the strap, one is assuming that the fiber output pulse is the convolution of the transmitter output pulse shape, the fiber impulse response, and the impulse response of the receiver, as would be predicted by (3.5.1). If the fiber under test has an impulse response whose rms width is narrow compared to the overall pulse response rms width obtained with the strap, then the deconvolution process (implemented by dividing the transforms) is subject to large errors, due to system non-linearities. Often, to obtain maximum sensitivity, avalanche photodiodes are used in the receiver. It is doubtful whether such detectors are linear enough to allow significant amounts of deconvolution. To avoid this problem, one tries to use sources, detectors, and receivers which are fast enough so that the output pulse measured with the fiber under test has an rms width which is dominated by the rms width of the fiber impulse response. To minimize nonlinearity, one often adds fixed optical attenuation to the optical path during the strap measurement, comparable to the attenuation of the fiber under test.

Another source of error in the deconvolution process is trigger time jitter introduced in the trigger delay generator. It is relatively straightforward to show that the averaged receiver output pulse shape obtained with trigger jitter present is the convolution of the averaged output pulse shape which would be observed without jitter; and the probability density function of the jitter. The reference deconvolution process (normalizing by the strap measurement) removes this jitter effect. However, as mentioned above, nonlinearities limit the usefulness of deconvolution. Also, the jitter probability density may be a function of the amount of trigger delay being set on the delay generator. Thus the jitter probability density function may be different in the fiber under test and strap measurements. Noise can be removed by averaging more and more samples of the pulse response. However, this increases the measurement time and may increase the vulnerability of the measurement to drifts of various sorts in the equipment.

Pickup should be minimized by careful system design. Residual pickup can be subtracted out by averaging the receiver response with the optical source disconnected from the fiber (but with the receiver still being triggered by the transmitter and the delay generator).

Pulsed lasers often exhibit pulse-to-pulse amplitude fluctuations. These fluctuations can be normalized by dividing the averages of the samples of the fiber pulse response vs. time by the averages of samples of the response taken at a fixed point in time, or by using a beam splitter to sample and average the transmitter output at the fiber input.

Pulsed lasers tend to have a broad spectral content. This increases the measured fiber impulse response width because of material dispersion (variation in delay vs. wavelength). If the laser has a spectral width of, say, 4 nm (typical), then at 820-nm wavelength this results in a material dispersion contribution to the impulse response width of about 0.4 ns km^{-1}. The material dispersion contribution to the rms width adds to the modal delay spread contribution to the rms width as the square root of the sum of the squares (as will be discussed further in Chapter 4). Therefore we can neglect the error caused by material dispersion effects if the total rms width of the fiber impulse response is more than about 1.5 ns km^{-1}. For better fibers, this effect must be taken into account.

In order to improve the temporal resolution of the measuring set, one can use the "shuttle pulse"[27] approach shown in Figure 3.55. Here partially silvered mirrors at the ends of the fiber under test allow the transmitted pulses to pass several times through the fiber. The added attenuation in this technique (for multiple passes) is traded off against the increased pulse spreading associated with the multiple passes. If one assumes that there is no mode mixing in the fiber or at the mirrors, then, for example, a "third-pass" pulse response simulates a fiber whose impulse response is scaled in time from the actual fiber response by a factor of 3. Thus the rms

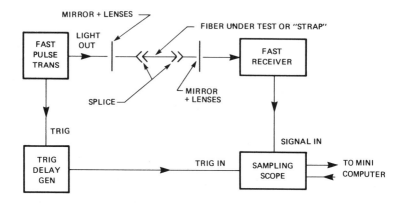

Figure 3.55. Shuttle pulse measuring set.

width associated with the third-pass response would be three times the rms width of the (single-pass) impulse response of the fiber. Figure 3.56 shows a sequence of pulse responses observed on a shuttle pulse system for a 0.1-km-long fiber.

As an alternative to time domain measurements, one can measure transfer functions directly in the frequency domain. Figure 3.57 shows such a measurement system using, in this case, a broad-band incoherent source. The output of the incoherent source is band limited by an optical filter (mono-

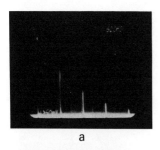

a

Figure 3.56. Actual shuttle pulse outputs. (a) Composite of 1, 3, 5, 7, 9th pass through a 106 meter long fiber (500 ns/div). (b) Individual responses (lns/div) left-to-right: 1st pulse ($L = 106$ m), 5th pulse ($L = 954$ m), and 10th pulse ($L = 2014$ m). (Courtesy of L. G. Cohen, Bell Laboratories.)

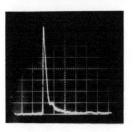
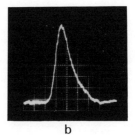
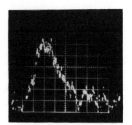

b

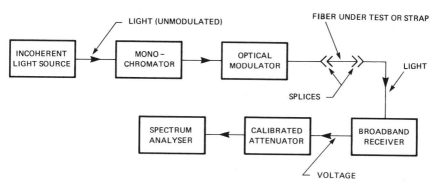

Figure 3.57. Transfer function measuring set.

chromator) and passes through an electro-optic modulator. The modulator output power varies sinusoidally above and below a bias level. The modulator output is launched into a fiber under test or a strap fiber, and propagates to a detector–receiver. The modulation frequency is varied while the receiver output level is monitored. One thus records the receiver output vs. modulation frequency for the fiber under test and for the strap. By dividing these two measurements one obtains the transfer function of the fiber. Some advantages of this method compared to the others described above are less nonlinearity problems than with the pulse method, more control of the source spectral width, less digital processing, and more flexibility in the choice of the source and the measurement wavelength. Disadvantages are the need for an external modulator (for this particular type of source) and less direct time domain information (it is desirable to see the pulse response in "real time" to understand what the system is doing). When making this measurement one must be careful that the source is sufficiently optically band limited so that all components entering the electro-optic modulator are modulated in phase (the bias point of the electrooptic modulator is very wavelength sensitive). Otherwise, chirping and dechirping effects, as described in Section 3.3.1 above, can occur.

As with the measurement of loss, the measurement of impulse response, rms width, or bandwidth is very sensitive to the launch conditions. Reproducible results can be obtained using incoherent sources and/or mode scramblers.

As mentioned above, material dispersion is an important factor when sources with broad spectral widths are being used. A simple way to measure the material dispersion of a fiber is to record the delay through the fiber using narrow pulses from lasers at two different wavelengths. The material dispersion (ns nm^{-1} km^{-1}) is obtained by dividing the delay difference by the wavelength difference of the lasers and normalizing to the fiber length.

Problems

1. The rms spectral width of an LED source is 50 nm. A fiber has a material dispersion at the center wavelength of the source of 100 ps km^{-1} nm^{-1}. Calculate the rms pulse spreading in ns km^{-1} due to material dispersion.

2. For the circuit of Figure 3.4, assume that the TTL gate output voltage is either 0 or 5 V. Assume that the base emitter voltage drop of the driver transistor is 0.7 V. Plot the required emitter-to-ground resistance value vs. LED current for currents between 10 and 200 mA (neglect transistor-emitter resistance and assume transistor $\beta \gg 10$).

3. Plot N_B of (3.4.1) for frequencies f between 100 MHz and 10^{14} Hz for $T = 300$ K, $T = 10,000$ K.

4. For an avalanche detector, prove $\langle m^2 \rangle / M^2 \geq 1$ [see equation (3.4.33)].

5. Verify that $P_{M,x}(m)$ of (3.4.37) satisfies (3.4.36).

6. Verify that (3.5.22) follows exactly from (3.5.21).

7. A 50-Mb s^{-1} transimpedance digital system receiver has an average optical signal input level of -55 dBm. The transimpedance is 4000 Ω. The detector responsivity is 0.5 A/W $\times$ detector gain M. Calculate the receiver output in millivolts for $M = 50$. Calculate the number of incident photons per pulse for $hf = 2 \times 10^{-19}$ assuming half the input pulses are present and half are not.

8. Using the Gaussian approximation, calculate and plot the receiver sensitivity (dBm) vs. avalanche gain for a digital receiver with a required error probability of 10^{-9}, $Z = 1500$, Ext $= 0$, $k = 0.03$, $hf = 2 \times 10^{-19}$, $B = 5 \times 10^7$ bits sec^{-1}. What is the optimal gain M? What is the associated minimum number of photons per pulse?

9. For an analog receiver incorporating on APD, $Z = 1500$, $k_m = 0.5$, $hf = 2 \times 10^{-19}$, $k = 0.03$, plot P_0 (in dBm) vs. M for SNR $= 10^5$ and $B = 5 \times 10^6$. Repeat for SNR $= 10^4$, SNR $= 10^6$.

Systems

In this chapter we shall consider systems made up of the subsystems discussed in Chapter 3. First, we shall investigate the interactions and trade-offs between subsystem parameters. Then we shall look at examples of hypothetical and actual fiber optic systems. We shall also compare these to their copper cable counterparts.

4.1. Interactions between Subsystems

In the discussions in Chapter 3 there were several aspects of subsystems which we deferred to this chapter because they involved interactions between the transmitter, the fiber, and the receiver. We will now discuss these aspects in some detail.

4.1.1. Modes and Coupling [1, 2]

As was mentioned several times in Chapter 3, typical multimode fibers can support several hundred independent, low-loss propagating modes. In order to understand the meaning of modes, it is helpful to consider a two-dimensional propagation situation as depicted in Figure 4.1. What is shown is a one-dimensional aperture illuminated by a plane wave. The plane wave emerges from the aperture as shown, diverging with an angle λ/D, where λ is the wavelength of the illumination and D is the width of the aperture. This divergence phenomenon is called "diffraction," and can be derived from standard field theory. In order to understand the origin of the divergence angle, λ/D, imagine that the aperture is illuminated by an arbitrary field.

The one-dimensional field in the aperture (from $x = 0$ to $x = D$)

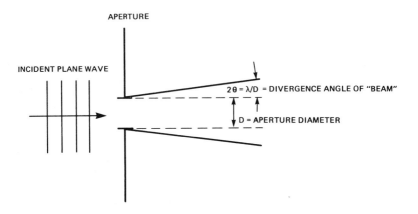

Figure 4.1. Two-dimensional emission from an aperture, $k = 0$ term.

could be expanded in a spatial Fourier series as a function of the spatial parameter x as follows:

$$\varepsilon(x) = \sum_{k=-\infty}^{\infty} a_k e^{i2\pi kx/D} \tag{4.1.1}$$

where D is the aperture width.

The lowest-order term in the expansion ($k = 0$) corresponds to a constant field across the aperture, and leads to the emission already shown in Figure 4.1. The term with $k = 1$ corresponds to a field which undergoes 2π rad of phase shift from $x = 0$ to $x = D$. This field corresponds to the emission shown in Figure 4.2 where if the emission angle relative to the normal is λ/D then the phase delay between the top and the bottom of the aperture is exactly 2π rad as required. Thus if an arbitrary field in the aperture is expanded in a spatial Fourier series, each term, k, in the expansion represents a diverging plane wave with an angle approximately $k\lambda/D$ relative to the normal to the aperture. The angular spacing between these "modes" is λ/D. Since the total field occupied by these modes must fill all of space, each diverging plane wave must have a divergence angle of λ/D.

For a two-dimensional circular source of diameter D radiating into a solid angle Ω_s the number of modes being radiated is equal to the total solid angle of the radiation, divided by the solid angle occupied by a single mode. This is given by

$$N_s = \text{No. of modes radiated} = \frac{\Omega_s}{(\pi/4)(\lambda/D)^2} = \frac{4}{\pi\lambda^2}(\Omega_s D^2) \cong \Omega_s A_s/\lambda^2 \tag{4.1.2}$$

where A_s is the source area and Ω_s is the source solid angle of emission.

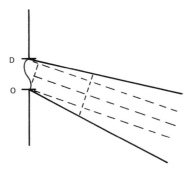

Table 4.2. Two-dimensional emission from an aperture, $k = 1$ term.

For example, consider a light-emitting diode with a total emitted power P_s (watts). As we shall see, the power emitted by the source is not, in itself, representative of the power which can be launched into a fiber. What is representative is the source brightness, B_s, defined as the power emitted per unit source area per unit solid angle of emission:

$$B_s = P_s/\Omega_s A_s \qquad \text{W}/(\text{cm}^2 \cdot \text{sr}) \qquad (4.1.3)$$

where P_s is the total source output in watts, A_s is the source area in cm^2, and Ω_s is the source emission solid angle in steradians.

From (4.1.2) and (4.1.3) we see that the radiated power per mode is given by the brightness multiplied by the square of the nominal source wavelength:

$$P_s/N_s = P_s/(A_s\Omega_s/\lambda^2) = B_s\lambda^2 \qquad \text{W}/\text{mode} \qquad (4.1.4)$$

It is also interesting to normalize the power per spatial mode by the optical bandwidth of the source to obtain a quantity one could call the energy per spatial–temporal mode

$$P_s/N_sBW_s = B_s\lambda^2/BW_s \qquad \text{W}/(\text{sr} \cdot \text{Hz}) \qquad (4.1.5)$$

where BW_s is the source bandwidth in hertz.

A typical "high-brightness" 850-nm GaAs LED would have a brightness of 100 W sr^{-1} cm^{-2}. Its power in watts per mode would be $100 \times (0.8 \times 10^{-4})^2 = 0.64$ μW/mode. Such a source would typically have a bandwidth of about 10^{13} Hz at room temperature, or from (4.1.5) an energy of about 0.64×10^{-19} J per spatial–temporal mode. (It is noteworthy that the most powerful LEDs have an energy per spatial–temporal mode comparable to that of a single photon. This is not a coincidence, but follows from the fact that with more than one photon per spatial–temporal mode, the LED-active region would begin to have gain rather than loss, and lasing action would begin.)

The amount of power that can be coupled into a fiber from an LED is limited by the product of the LED power per mode and the number of fiber modes. Thus a fiber with a few hundred modes of propagation could capture ideally about 100 μW from the above LED. If the LED emits 1 mW of total power, then the coupling efficiency in this case is at best 10%.

The above ideal coupling efficiency cannot be improved upon by the use of lenses, mirrors, tapered fibers, etc. To improve this coupling efficiency would violate what is known as the "law of brightness." If the source emits into 1000 modes independently (noncoherently as in the LED case) and the fiber propagates 100 modes, then the maximum coupling efficiency is 10%. On the positive side, this implies that there are ten independent positions the fiber can have relative to the source, and still achieve the maximum theoretical coupling. Thus although the coupling is not efficient, it is also not critical.

It is possible that a source emitting in, say, 1000 modes couples less than 10% of its output into a fiber with, say, 100 propagating modes. This may occur because of a mismatch of the fiber and source modes, or difficulty in bringing the fiber sufficiently close to the source. For example, if the source area is smaller than the area of the fiber core, then one might require a simple lens melted on the fiber end to achieve maximum coupling efficiency as shown in Figure 4.3. Once again, the lens only helps to achieve a coupling efficiency near the theoretical maximum (ratio of fiber modes to source modes), by trading area against solid angle. No further increase is possible.

A typical injection laser source emits light in one of several spatial modes. Thus in principle a fiber with several hundred propagating modes should be able to capture nearly all of the light emitted by a laser. Indeed

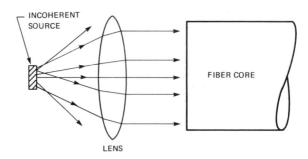

Figure 4.3. Improving coupling with a matching lens.

laser-to-multimode fiber coupling efficiencies in excess of 50% are typical. To achieve this excellent coupling, the laser modes must be matched to the fiber modes. Figure 4.4 shows how a laser emitting a highly divergent mode may be coupled to a fiber whose N.A. is inadequate to capture that mode. The lens collimates the laser output, making the laser look like a larger-area, smaller-divergence-angle emitter. Again, the lens matches source modes to fiber modes—it cannot couple several source modes simultaneously into fewer fiber modes.

There is a strategy whereby the output of an LED could be more efficiently converted into light propagating in a fiber, without violating the law of brightness. One could use the LED to optically "pump" a laser (e.g., a miniature Nd:YAG fiber laser). The laser light produced could then be coupled into a fiber. This scheme circumvents the law of brightness since nonlinear conversion of incoherent to coherent light has taken place. Miniature LED-pumped lasers have been built in the laboratory, but are not used at present except for experimental applications.

The constraints on splicing, coupling fibers to fibers, are the same as for coupling sources to fibers. A fiber with 110 propagating modes (uniformly excited by an incoherent source) can only couple about 90% of its light into a fiber with 100 propagating modes. If the splice is not aligned properly or if there is a mismatch in the modes (e.g., a large-core, small-N.A. fiber coupling into a small-core, large-N.A. fiber), then the coupling efficiency could be much less. Lenses and tapers can help the mismatch problem by converting modes, but the limitations of the mismatch in the number of modes cannot be circumvented. One should not confuse modes which are coherently excited, with modes which are independently incoherently excited. In principal a multimode fiber excited by a single-mode source can couple all of its output back into a single-mode fiber. Practically

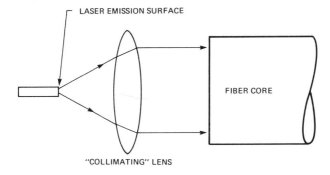

Figure 4.4. Coupling a laser to a fiber.

speaking, however, once the multiplicity of modes have undergone propagation for a finite distance, the phase differences of the different modes would make such a coupling nearly impossible to implement. Thus as a practical matter, the mode number ratio limits the coupling efficiency between fibers even for coherent excitation.

Of all the coupling problems, coupling the fiber to the detector is the least problematic. Typically the detector area is much larger than that of the fiber core. Thus from a geometrical point of view, it is easy for the detector to capture all of the light emitted from the fiber, provided that the fiber is reasonably close to the detector surface. The biggest limitation on coupling here is the reflection at the interface between the fiber and the detector. This can be minimized by the use of "index-matching" material and anti-reflection coatings. One may ask if there is any motivation to reduce the size of the detector. In general, the only motivating factor is detector capacitance (and perhaps leakage). Fortunately very-low-capacitance detector chips can be fabricated from silicon, which are still large in area compared to the fiber core. In most applications, the capacitance of the detector "header" (package) and the preamplifier input capacitance dominate the capacitance of the detector.

4.1.2. Noise from Source Fluctuations [3, 4]

In Chapter 3 we discussed noise introduced at the receiver because of the statistics of the detection process (quantum noise and avalanche gain statistics) and thermal noise from the preamplifier. There are several additional noises which appear at the receiver output that are due to fluctuations at the transmitter or due to interactions between the transmitter and the launch-sensitive characteristics of the fiber.

There is a source of noise associated with light-emitting diodes called "beat noise." Each spatial mode emitted by an LED has an amplitude which fluctuates in time in a random fashion. The field amplitude of each mode can be modeled by what is called a complex Gaussian random process. These mode amplitudes are statistically independent random processes with typical correlation times on the order of 10^{-13} s. Let us consider what happens when light originating from an LED illuminates an ideal photodetector to produce an average current plus noise. The noise at the detector output consists of quantum noise plus noise due to the random fluctuations of the incoherent LED emission. That is, if the total average power incident on the detector is p_r watts, then the output current is given by

$$i_{\text{out}}(t) = ep_r/hf + i_{nq}(t) + i_{nb}(t) \tag{4.1.6}$$

where e is the electron charge, hf the energy in a photon, p_r the detected power, $i_{nq}(t)$ is the quantum noise, and $i_{nb}(t)$ the beat noise due to the random nature of the LED output signal.

If the noise is band limited to a bandwidth B, then the mean squared quantum noise is given by[3]

$$\langle i_{nq}^2(t) \rangle = 2(e^2 p_r / hf) B \qquad (\text{A}^2) \qquad (4.1.7)$$

The mean squared beat noise is given by

$$\langle i_b^2(t) \rangle = 2(e p_r / hf)^2 (B/bw_r)(1/N_r) \qquad (4.1.8)$$

where bw_r is the optical bandwidth of the received power and N_r is the number of spatial modes in the received optical illumination.

Comparing (4.1.8) to (4.1.7) we see that beat noise is negligible compared to quantum noise *unless* the received illumination satisfies

$$\frac{p_r}{N_r bw_r} > hf \qquad (4.1.9)$$

We recognize the left side of (4.1.9) as the received power per mode per unit bandwidth, i.e., the received energy per "space-time mode". From the discussion in Section 4.1.1 we know that even at the source itself, an energy per space-time mode near hf implies that the LED is near the point of gain and lasing. In most applications, the receiver is not likely to be illuminated directly by the source (i.e., there will be intervening attenuation) and the LED is not going to be near the lasing point in its energy per space-time mode. Thus for most practical applications one can neglect beat noise compared to the ubiquitous quantum noise.

When laser sources are used, there are two sources of noise associated with laser instabilities. The first source is closely related to the beat noise described above. Equations (4.1.7)–(4.1.9) show that beat noise in LEDs is negligible unless the LED is near the point of lasing. On the other hand, a laser slightly above threshold is just such a source. Therefore, if one observes the output of an injection laser with a photodiode (assuming little attenuation between the laser and the detector), one will see a rapid increase in noise (above the expected quantum noise level) as the laser drive current passes from below threshold to just above threshold. Further increases in the drive current will cause the laser to enter its normal saturated mode of operation, where these excessive beat noise contributions should disappear.

A second source of laser-related noise is associated with spatial mode instabilities.[4] Generally, injection lasers will oscillate in more than one spatial mode. These modes may be present simultaneously, or the laser may "hop" from mode to mode randomly in time. The output from such a source can be captured by a multimode fiber with very high coupling ef-

ficiency. However, because of the different attenuations and delays associated with different fiber modes, the output of the fiber can fluctuate as the
laser mode output pattern changes. This can introduce receiver output
noise which may be important in analog systems operating at high signal-
to-noise ratios (or even in digital systems if severe enough). Since this noise
is not fundamental, and involves interactions between imperfect lasers and
imperfect fibers, it is difficult to characterize quantitatively. Experience has
shown this noise to be negligible in actual digital system installations, but
important (and unpredictable) in analog system applications.

4.1.3. Delay Distortion [5]

Delay distortion as mentioned in the earlier chapters refers to the
spreading in time of a narrow pulse as it propagates along a fiber. There are
two sources of delay distortion: modal delay spread and chromatic dispersion. Modal delay spread refers to the differences in the group delays
of the various modes in a multimode fiber. Chromatic dispersion refers
to the variation of the group delay in a given mode, as a function of the
wavelength exciting that mode.

Consider first a coherent source, i.e., a laser, producing a narrow output
pulse, $h_p(t)$, of power (watts). Let this pulse illuminate the input of a fiber
whose impulse response is $h_{\text{fiber}}(t)$. From the power linearity approximation
given in Chapter 3, the fiber output pulse is

$$h_{\text{fiber out}}(t) = \int h_p(t') h_{\text{fiber}}(t - t') dt' \qquad (4.1.10)$$

Regardless of the exact shape of $h_p(t)$ and $h_{\text{fiber}}(t)$ it is straightforward to
show, from the definition of the rms width, that

$$\sigma_{\text{out}}^2 = \sigma_p^2 + \sigma_f^2 \qquad (4.1.11)$$

where σ_{out} is the rms width of the output pulse, σ_p the rms width of the input
power pulse, and σ_f the rms width of the fiber impulse response. In equations (4.1.10) and (4.1.11) we have assumed that $h_{\text{fiber}}(t)$ represents the modal
delay spread. We have not yet considered chromatic dispersion effects.

Next let the fiber impulse response at a given wavelength, λ, be defined as $h_{\text{fiber}}(\lambda, t)$. Suppose the source produces an output spread over a
band of wavelengths, and let that portion of the total output at a particular
wavelength λ have the form $h_p(\lambda, t) d\lambda$. Here $d\lambda$ is an increment of wavelength.

The total source output is given by adding the outputs in the increments $d\lambda$. That is,

$$h_p(t) = \int h_p(\lambda, t)\, d\lambda \qquad (4.1.12)$$

If this source illuminates a fiber, the output produced by the increment of input power in the spectral band $d\lambda$ is

$$h_{\text{out}}(\lambda, t)\, d\lambda = \left[\int h_p(\lambda, t')\, h_{\text{fiber}}(\lambda, t - t')\, dt' \right] d\lambda \qquad (4.1.13)$$

The total fiber output is therefore given by

$$h_{\text{out}}(t) = \int h_p(\lambda, t')\, h_{\text{fiber}}(\lambda, t - t')\, dt'\, d\lambda \qquad (4.1.14)$$

We can simplify (4.1.14) somewhat by making some assumptions which are generally valid. First assume that as the wavelength, λ, changes, the fiber impulse response is simply delayed in time owing to the overall variation of group delay with wavelength. That is,

$$h_{\text{fiber}}(\lambda, t) \cong h_{\text{fiber}}(\lambda_0, t - D(\lambda - \lambda_0)) \qquad (4.1.15)$$

where λ_0 is the nominal wavelength of the source, D = a delay per unit wavelength = δL, L is the fiber length, and δ is the delay per unit wavelength per unit distance. Also assume that the shape of the source output pulse in an increment of wavelength, $d\lambda$, is approximately independent of λ. That is,

$$h_p(\lambda, t) = \tilde{h}_p(\lambda_0, t)\, S(\lambda) \qquad (4.1.15a)$$

where

$$S(\lambda) = \int h_p(\lambda, t)\, dt, \qquad \tilde{h}_p(\lambda_0, t) = h_p(\lambda_0, t)/S(\lambda_0),$$

and where λ_0 is the source nominal wavelength. Then (4.1.14) simplifies to

$$h_{\text{out}}(t) = \int \tilde{h}_p(\lambda_0, t')\, h_{\text{fiber}}(\lambda_0, t - t' - D(\lambda - \lambda_0))\, S(\lambda)\, d\lambda\, dt' \qquad (4.1.15b)$$

$$= \int \tilde{h}_p(\lambda_0, t')\, h_{\text{fiber}}(t - t')\, dt'$$

where

$$h_{\text{fiber}}(t) = \int h_{\text{fiber}}(\lambda_0, t - D(\lambda - \lambda_0))\, S(\lambda)\, d\lambda \qquad (4.1.15c)$$

Let the source rms width, Λ_s, be defined as

$$\Lambda_s^2 = \int (\lambda - \lambda_0)^2 \, S(\lambda) \, d\lambda \left/ \int S(\lambda) \, d\lambda \right. \qquad (4.1.16)$$

where

$$\lambda_0 = \int \lambda S(\lambda) \, d\lambda \left/ \int S(\lambda) \, d\lambda \right.$$

Let the rms width of the fiber impulse response $h_{\text{fiber}}(\lambda, t)$ at a fixed wavelength be defined as σ_{fm}. Then the rms width of the overall fiber impulse response $h_{\text{fiber}}(t)$ defined in (4.1.15c) is

$$\sigma_f^2 = \sigma_{fm}^2 + \sigma_{fd}^2 \qquad (4.1.16a)$$

where

$$\sigma_{fd}^2 = D^2 \Lambda_s^2 = \delta^2 L^2 \Lambda_s^2$$

Thus the rms width of the output pulse is given by

$$\sigma_{\text{out}}^2 = \sigma_f^2 + \sigma_p^2 = \sigma_p^2 + \sigma_{fm}^2 + \delta^2 L^2 \Lambda_s^2 \qquad (4.1.16b)$$

where σ_p is the rms width of the optical input power pulse (assumed independent of wavelength), σ_{fm} is the rms width of the fiber impulse response at a fixed wavelength (due to mode delay spread) assumed independent of wavelength, δ is the chromatic dispersion of fiber in nS (nm^{-1} km^{-1}), typically 0.1 at 0.8-μm wavelength, L is the fiber length in kilometers, and Λ_s is the source rms spectral width in nanometers, typically 17–20 nm for GaAs LEDs and less than 1 nm for GaAs injection lasers. It is important to note that the modal delay spread and chromatic dispersion effects contribute to the overall fiber rms impulse response width as the square root of the sum of the squares. Also the chromatic dispersion contribution to the rms width of the fiber impulse response is proportional to the fiber length. Recall from Chapter 3 that owing to variations in the loss amongst the modes of the fiber and owing to mode-mixing effects the modal delay spread contribution, σ_{fm}, can increase less than linearly with the fiber length.

4.1.4. Tradeoffs of Coupling Efficiency, Loss, and Delay Distortion [6]

In many fiber system applications one can make a tradeoff amongst coupling efficiency between the source and the fiber, fiber loss (attenuation), and delay distortion in the fiber. At the receiver, it is generally desirable to have as large a signal as possible illuminate the photodetector. Thus one would desire a high coupling efficiency between the transmitter source and the fiber, and one would desire as low a loss in the fiber as possible. However, the receiver sensitivity is lower if the fiber introduces too much pulse spreading or equivalently too much band limiting due to delay distortion. One can speak of a system being loss limited or delay distortion limited. A loss-limited system is one in which the maximum fiber spacing between the transmitter and the receiver is limited by fiber loss. A delay-distortion-limited system has a transmitter-to-receiver spacing limited by pulse spreading or band limiting in the fiber. Ideally to maximize the spacing, one would trade off received power against delay distortion. In many practical circumstances one cannot optimize on a case by case basis. Therefore, when picking standards for fibers, one tries to anticipate most common applications and one tries to design fibers which are a reasonable compromise for the varying needs of those applications. It is interesting to identify some of the tradeoffs.

For a coherent (laser) source, the amount of light one can couple into a fiber is for the most part independent of the fiber parameters, provided the fiber can transmit several modes. The larger the fiber core area and the larger the fiber numerical aperture, the easier the coupling problem becomes. However, in principle as long as the number of fiber modes exceeds the number of possible source modes, complete coupling is possible.

For LED sources, the number of source modes is generally one or two orders of magnitude more than the number of fiber modes. The amount of power that can be coupled into the fiber is proportional to the square of the numerical aperture and is also proportional to the fiber area. Thus with an LED source whose emission area is comparable to the fiber core area, one would like the fiber numerical aperture to be as large as possible to maximize the coupling efficiency.

Whether one is dealing with a step index fiber or a practical graded index fiber, the pulse spreading associated with modal delay spread is proportional to the fiber index step, Δ. Therefore the pulse spreading is proportional to the square of the fiber numerical aperture $[N.A. = n(2\Delta)^{1/2}]$. Thus to reduce pulse spreading in transmission one would like to reduce the numerical aperture (which works against coupling efficiency with an LED source).

The portion of the pulse spreading associated with chromatic dispersion can be reduced by limiting the optical bandwidth of the source. With an LED this can be accomplished by using an optical filter between the source and the fiber. Such a filter will inevitably have some insertion loss, and at the minimum, if the filter reduces the LED spectral width by a factor of 2 it must introduce 3 dB of loss. Thus one can trade off chromatic dispersion against coupling efficiency by using an optical filter.

When fibers are made, the numerical aperture is increased by adding germanium, boron, phosphorous, and other materials to the basic silica which comprises the glass. The presence of these materials increases the scattering loss in general. Thus highly doped, large-numerical-aperture fibers tend to have more scattering loss. On the other hand, larger-N.A. fibers are easier to couple into, easier to splice, and radiate less at bends.

Fibers with a large core area are easier to couple into and easier to splice; however, they are more costly (more material), more subject to breakage at bends, and they radiate more at bends.

Fibers with mode mixing are more lossy, but they have less delay distortion, as discussed in Chapter 3.

Thus we see that the optimization of fibers is a complex problem involving numerous tradeoffs. Remarkably, independent attempts to arrive at a standard fiber have led to fairly consistent results. For long-distance transmission where loss is critical, typical fibers have a core diameter of about 50–75 μm and a numerical aperture of 0.15–0.20. There is some incentive to use a different fiber for short-distance applications where LED–fiber coupling efficiency and ease of splicing are more critical than loss. Here larger core areas and larger N.A.s are chosen.

One typically finds that intentionally increasing mode mixing to reduce delay distortion is a costly approach in terms of loss. It is possible to introduce carefully controlled mode mixing without coupling guided modes to radiation modes. However, as a practical matter this is difficult to do. The trend is to minimize mode mixing along with minimizing other sources of loss. To keep delay distortion under control one can use carefully index-graded fibers, or perhaps in the future, single-mode fibers.

4.1.5. Low-Frequency Delay Distortion — Tails

As pointed out in Chapter 3, Section 3.5.1, the fiber impulse response can include a "tail" as shown in Figure 3.47. This tail produces a "bump" in the transfer function of the fiber as shown in Figure 3.48. The impact of such a tail is negligible if the fiber system receiver is limited by thermal noise, and if the bump rolls off at a frequency which is low compared to the overall receiver bandwidth being utilized. The bump can then be removed by ac-coupling the receiver or possibly by an equalizer which compensates with low frequency loss. If quantum noise is an important factor in the receiver (e.g., with an APD receiver), then the average optical power at the receiver input associated with the low-frequency bump will add to this quantum noise without significantly contributing to the signal portion of the receiver output. Thus the low-frequency bump is equivalent to an unmodulated bias offset in the received optical signal.

The presence of such a bump also affects the definition of the fiber impulse response rms width, and the fiber bandwidth. If one assumes that the bump will be ac coupled or equalized out, then the fiber bandwidth can be defined by extrapolating the frequency response toward zero from a frequency beyond the bump. In the time domain, the rms width of the impulse response can be similarly defined by ignoring the tail.

4.2. Examples of Digital Systems (7-10)

Having derived and presented the material in Chapter 3 on subsystems and the above material on interactions between subsystems, we will now consider some actual system examples.

4.2.1. A 10-Mbaud Binary Short-Haul Trunk Carrier System

In telephone jargon, a "short-haul trunk carrier system" is a transmission "facility" between telephone buildings, a few miles or tens of miles apart, which carries several tens or hundreds of telephone voice circuits simultaneously. A typical digital trunk carrier system would convert standard 4-kHz voice signals to a multiplexed digital form using the equipment shown in Figure 4.5. The "channel bank" converts 24 voice signals to digital form and combines them to produce a binary data stream with a clock rate of 1.544 MHz. Thus each voice channel occupies 64 kbaud out of the total. These 1.544-Mbaud signals, called DS1 signals (digital signal 1), can be "multiplexed" together to produce a composite signal at a higher data rate. For example, 6 DS1 signals could be interleaved to produce a composite signal at about 10 Mbaud (including extra "bits" for control and maintenance purposes). The function of the transmission system is to carry this data stream to a distant location without introducing errors.

Figure 4.6 shows a typical fiber "short-haul trunk" transmission system. At the input to the system a scrambler is used to randomize the binary signal. That is, the scrambler removes long sequences of "zeros" and assures a frequency of transitions between "ones" and "zeros" which is representative of random data. The scrambler output (data and clock) drives an optical transmitter which produces pulses of power at the clock rate. The shape of

Figure 4.5. Channel bank.

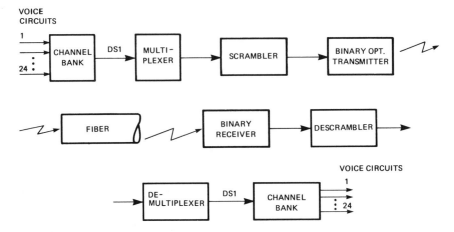

Figure 4.6. Typical fiber short-haul trunk transmission system (one way).

an optical pulse need not match the scrambler output pulse shape exactly. That is, the scrambler data output could consist of half-duty-cycle pulses, while the transmitted optical pulses could be full duty cycle. The optimal transmitted pulse shape will be discussed below. The transmitter output is coupled to a fiber with some coupling loss. The coupled optical signal (power) propagates along the fiber, incurring attenuation and delay distortion. The resulting fiber output illuminates a receiver which amplifies, reshapes, and samples the detector output to produce a "regenerated" digital data stream and a clock output. These drive the descrambler, the demultiplexer, and finally the channel bank.

We shall assume that the transmitter incorporates a light-emitting diode. Typical transmitter specifications were given in Chapter 3. One parameter which is not discussed in detail was the output pulse width. At this data rate, the LED will be peak power limited. Thus, the wider the optical pulse width, the more average power launched into the fiber. On the other hand, the wider the transmitted pulse width, the larger will be the rms pulse width at the receiver input, and the less the receiver sensitivity. The exact loss in receiver sensitivity as a function of the transmitted pulse width depends upon the other contributor to the received rms pulse width, fiber delay distortion, and on the exact shape of the received pulse. However, the loss in receiver sensitivity (sensitivity penalty) as a function of the received rms pulse width can be approximated crudely as follows (for penalties less than a few decibels)[12]:

$$\text{sensitivity penalty} \leq \frac{K_0 \sigma_r^2}{T^2} = \frac{K_0(\sigma_p^2 + \sigma_f^2)}{T^2} = \frac{K_0\left[(\Gamma T)^2/12 + \sigma_f^2\right]}{T^2} \quad (4.2.1)$$

where σ_r is the received pulse rms width, σ_p is the transmitted rms pulse width, σ_f is the rms width of fiber impulse response, T is the time slot width = 1/baud, Γ is the transmitted pulse duty cycle, and K_0 is a constant approximately equal to 20 in typical applications.

The average received optical power is proportional to the transmitted pulse duty cycle. The optimal duty cycle is obtained by maximizing the difference, Δ_R, between the received average optical power (in dBm) and the sensitivity penalty (in dB). That is,

$$\Delta_R = 10 \log \Gamma - K_0 \left[(\Gamma T)^2/12 + \sigma_f^2 \right]/T^2 \qquad (4.2.3)$$

if we set

$$\frac{\partial \Delta_R}{\partial \Gamma} = 0 = \frac{4.34}{\Gamma} - \frac{K_0 \Gamma}{6} \qquad (4.2.3a)$$

We obtain

$$\Gamma_{\text{optimal}} = (26/K_0)^{1/2} \qquad (4.2.4)$$

A value of $\Gamma = 1$ corresponds to full-duty-cycle (nonreturn-to-zero, NRZ) transmitted pulses. Thus from (4.2.4) we see that even with K_0 as large as 26, the optimal transmitted pulse width is full duty cycle. From a practical point of view, full-duty-cycle pulses are easier to produce because of transmitter electronics and source bandwidth limitations. Thus both theory and practical constraints lead to a choice of full-duty-cycle transmitted pulses.

Figure 4.7 shows a transmitter "card" for an actual 10-Mbaud LED

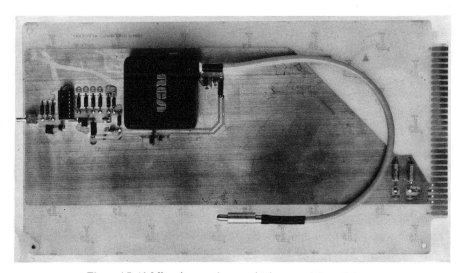

Figure 4.7. 10-Mbaud transmitter card (photograph by J. Schultz).

system. The square "module" contains an LED source and a driver designed to interface with a balanced ECL signal. A fiber pigtail is shown emerging from the module and terminating on a connector. The drive signal for this module is used directly to turn the source on and off. Thus no clock is required. A TTL-to-ECL converter is also shown on the card, since in this particular application, the interface to the scrambler output was TTL and the distance between the scrambler and the transmitter card was short.

A typical LED transmitter can couple about -15 to -18 dBm average power into a typical fiber with a 62.5-μm core and a 0.2 N.A. A typical avalanche photodiode receiver will require about 1000 photons per bit of average power (20 dB above the quantum limit) including some margin, for a 10^{-9} error rate. At 10-Mbaud and 0.82-μm wavelength this corresponds to

$$p_{\text{av min receiver}} = 1000 \times 2 \times 10^{-19} \times 10^{7}$$

$$= 2 \times 10^{-9}$$

$$= -57 \, dBm \tag{4.2.5}$$

Thus the allowable loss between the transmitter and the receiver, including fiber loss and splices and connectors, is about 40 dB. For fiber delay distortion to have a negligible effect on the receiver sensitivity (less than 1 dB) we require the rms pulse spreading due to fiber delay distortion to be less than about $1/4$ of a time slot. This corresponds to 25 ns at 10 Mbaud. If we use a graded-index fiber, we can assume that chromatic dispersion associated with the broad output spectrum of the LED will dominate delay distortion. The chromatic dispersion at 0.82-μm wavelength is about 0.1 ns nm^{-1} km^{-1}. Thus for a typical LED with a rms spectral width of 20 nm, the rms pulse spreading associated with chromatic dispersion is about 2 ns km^{-1}. This implies that the fiber length can be as long as 12.5 km before the system would be limited by delay distortion. In other words, with 40 dB of allowable loss, the system will be loss limited for average fiber losses (including splices and connectors) exceeding 3.5 dB km^{-1}. Since 3.5 dB km^{-1} including splices and connectors is a very low attenuation, we can assume that the system will be loss limited and that the received pulse is essentially an attenuated replica of the transmitted pulse.

Figure 4.8 shows the schematic diagram of an actual receiver for a 10-Mbaud fiber system. Figure 4.9 shows the receiver "card". The square module contains an APD and a transimpedance amplifier with a transimpedance of 5000 Ω. A typical specification for such a module was given in Chapter 3. The module includes a fiber pigtail, terminated on an optical connector. The automatic gain control reduces the avalanche gain and the AGC amplifier gain in response to increasing optical signals as described

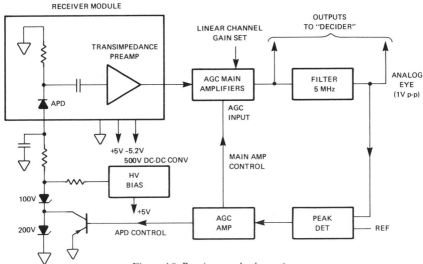

Figure 4.8. Receiver card schematic.

in Chapter 3. A dynamic range of 40 dB of optical signal level is obtained. Since the maximum loss between the transmitter and the receiver is 40 dB, this receiver is capable of automatically adjusting to any value of allowed loss from 0 to 40 dB.

Figure 4.10 shows a "decider card" which contains a phase-locked loop for clock recovery and a comparator sampler for regenerating the digital signal. Figures 4.11 and 4.12 show a typical fiber optics "bay" which inter-

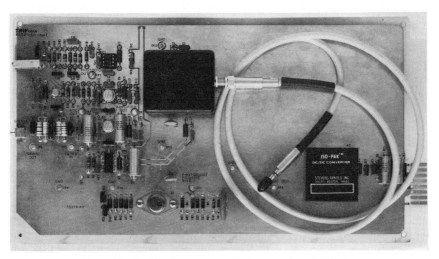

Figure 4.9. 10-Mbaud receiver card (photograph by J. Schultz).

Figure 4.10. 10-Mbaud decider card (photograph by J. Schultz).

Figure 4.11. Fiber optics shelves (photograph by J. Schultz).

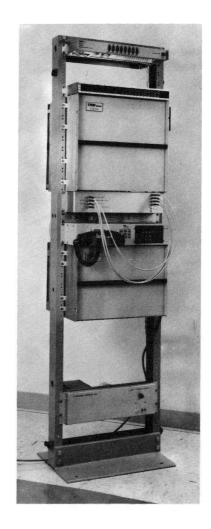

Figure 4.12. Fiber optics bay (photograph by
J Schultz).

faces up to 6 DS1 signals, including the multiplexer, main and standby
(redundant) fiber optic transmitters and receivers, a fiber optic crossconnect
panel, automatic main-to-standby transfer equipment, and power supplies
(-48-V input).

In order to operate the system over longer distances, it would be desirable to use a laser transmitter. The laser can launch perhaps 15–20 dB
more power into a typical transmission fiber and in addition does not produce significant amounts of delay distortion due to chromatic dispersion.
The problems associated with the complexity of laser drivers were described
in Chapter 3. Laser lifetime as a function of the device temperature in actual

operation is also a problem. Another factor, which involves the interaction of the laser and the receiver, is laser oscillation.

It has been observed that typical lasers will produce oscillating output pulses as shown in Figure 4.13. It would appear at first thought that such oscillations would be negligible if the oscillation frequency is above the cutoff of the receiver filters, typically near the baud. However, further thought will show that this is not the case. One way to see this is to consider the transmitted waveform to be the sum of a nonoscillating signal and a modulated oscillating signal as shown in Figure 4.13. The modulated oscillating signal has a spectrum as shown in Figure 4.14. We see that the modulation process generates side bands of the oscillation which can fall within the receiver baseband, even if the oscillation frequency is far above this base band. For negligible effect, it is necessary for the frequency of the oscillation to be 5–10 times the baud. Thus for a 10-Mbaud system, one would require the oscillation frequency of the laser to be no less than about 100 MHz.

Degradations due to laser oscillations are typical of numerous non-idealities which impair the sensitivity of the receiver. Other examples are

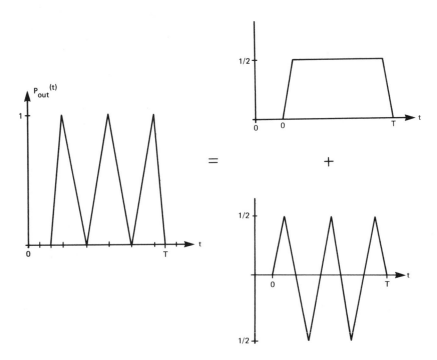

Figure 4.13. Oscillating laser output pulse.

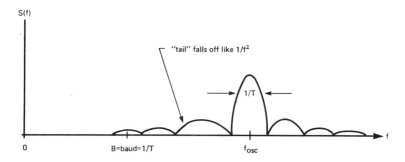

Figure 4.14. Spectrum of modulated oscillating signal (oscillation frequency $= f_{osc}$).

pulse-to-pulse amplitude and width variations, pulse overshoot, jitter in the rate of pulse arrival, deviations of the received pulse shape from nominal, deviations in the receiver filter characteristics from nominal, and power line noises. When designing a system allowance must be made for the cumulative effect of all of these degradations, either on a worst-case or on a statistical basis.

In designing a fiber optic trunk carrier system, it is interesting to note the constraints that fiber loss variations, fiber connector loss variations, source output variations and fiber length variations place on the "power budget" and the "dynamic range" requirements.

Figure 4.15 shows a typical telephone trunk end-to-end connection between a transmitter card and a receiver card. Connection No. 0 is between the transmitter source and the pigtail in the transmitter module. Since the transmitter power output is specified at the output of the connector at the other end of this pigtail, the coupling loss at this connection is relevant only in the amount it affects the tolerance specification at the pigtail output. Jumper 1–2 on the back of the "bay" interconnects the transmitter card pigtail with a "crossconnect panel". Jumper 2–3 interconnects the transmitter "appearance" on the front of the crossconnect panel with a fiber in the "fanout". The "fanout" splices to the fiber "riser" cable which runs between the "bay" and the "cable vault" in the basement of the telephone building. The outside plant cable splices to the riser in the cable vault. The outside plant cable is made up of sections perhaps 500 m to 1 km long which are spliced together in manholes. This cable enters the basement of the remote location where the above sequence of connections is reversed, finally terminating on the detector in the receiver module. We see from this figure that for a 4-km link as many as seven outside plant splices are required, along with two splices and three connectors at each end. If we allowed 0.5–2 dB per connector and 0.5–1 dB per splice we would have a

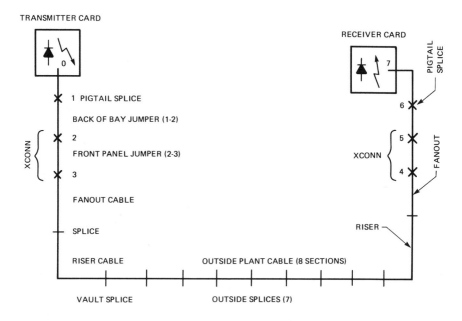

Figure 4.15. End-to-end optical connection.

total of 8.5–23 dB of loss in the splices and connectors depending upon the actual individual connection losses. This is a very sobering observation, particularly when one observes that the connection loss ranges given are fairly realistic for a real installation. If we are more optimistic, and assume that splice losses are 0.1–0.2 dB and connector losses are 0.1–1 dB we still obtain a total range of 1.7–8.2 dB for interconnection losses for a 4-km link. If the nominal fiber loss is 5 dB/km the actual fiber losses may range from 3 to 7 dB/km. Thus over a 4-km span, the fiber losses may range from 12 to 28 dB. Thus the total loss of a 4-km link could easily vary by ±11 dB from its nominal value of 25 dB. If one has an allowable loss of 40 dB between the transmitter and the receiver excluding receiver margin, then under worst-case conditions, the above 4-km link would allow 4 dB of additional margin for any transmitter output reduction from nominal with temperature and time, and for uncertainties in the exact length of the link. The receiver would have to be capable of absorbing the 22-dB range of fiber and connector losses plus this additional 4 dB of margin plus any allowance for the transmitter output being above nominal. Figure 4.16 shows the above power budget.

In order to bring the allowable nominal fiber and connector loss closer to the maximum allowable loss between the transmitter and the receiver,

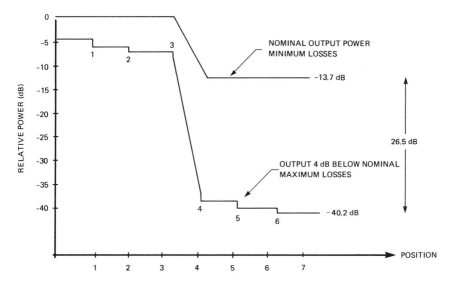

Figure 4.16. Power budget (excludes XMTR above normal output).

a tighter tolerance on fiber and connector losses is required. Tighter fiber loss tolerance can be achieved (for a price) by more control in the fiber fabrication and more selection during the fiber cable manufacture, and by mixing and matching fibers which are above and below nominal loss during the cable installations process (requires more costly installation practices). The variations in splice and connector losses can be reduced with tight control of the fiber parameters (e.g., core and cladding diameters, core–cladding concentricity, N.A.) and of the splice and connector installation.

4.2.2. Data Buses

The design considerations for a data bus are different from those for a telephone trunk carrier system because the distances to be spanned are much shorter and therefore the allowable cost of the end terminals is lower. Also in many present data bus applications, there is no fixed clock rate, that is, the signals are asynchronous. Figure 4.17 shows the block diagram of the transmitter of a typical point-to-point data bus. Figure 4.18 shows the block diagram of the receiver. Figure 4.19 shows the system specification sheet. The system uses an LED transmitter and a pin detector receiver. A three-level (differentiating) code is used (as described in Section 2.2.2) to accommodate the asynchronous data format with an ac-coupled transmission system. To minimize connector problems and to maximize the power which can be launched from an LED into the fiber, a relatively

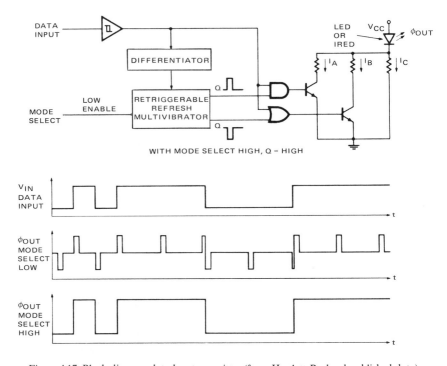

Figure 4.17. Block diagram data bus transmitter (from Hewlett Packard published data).

large-diameter, large-N.A. fiber is used. It is interesting to note the tradeoffs here. Since reliability and power consumption are critical, one is tempted to use a relatively low-radiance, low-current LED. To capture a reasonable amount of power, a large-core, large-N.A. fiber is required. The large core increases the fiber cost (more glass) and the large N.A. increases the loss and delay distortion. However, because the transmission distance is short, these factors are not as important as in the long-distance trunk.

In addition to short point-to-point links, there is considerable interest in fiber optic data bus networks with several access ports for transmitting and receiving. Figure 4.20 shows a "tree" structure consisting of a main terminal and several remote terminals. Figure 4.21 shows a "ring" structure. Data networks of this type are constrained by the considerable amount of overhead control bits required to maintain orderly communication amongst the several terminals. An advantage of fiber optics here is the ability to provide high data rates, allowing wasteful but simple data "protocols."

The largest challenge in the hardware (as opposed to systems and software) portion of a fiber data network is in the "terminal tap." There are basically two approaches to connecting a terminal to the network. In the

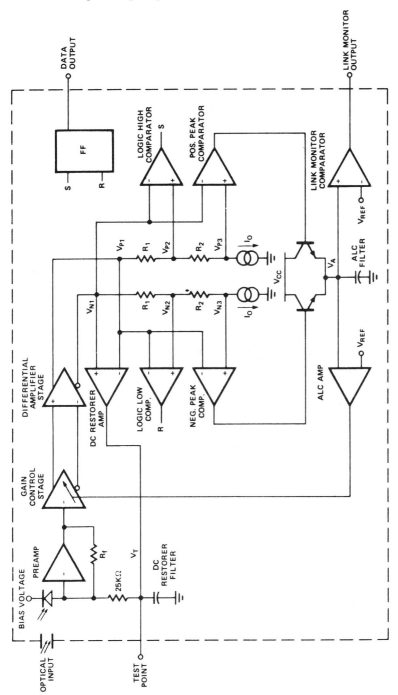

Figure 4.18. Block diagram of data bus receiver (from Hewlett Packard published data).

Sample Flux Budget Calculation

DATA SHEET PARAMETERS			MIN	TYP	MAX	UNITS	NOTES
HFBR-1002	Output Optical		50	100		μW	*
Transmitter	Flux		−13	−10		dBm	
HFBR-2001	Input Optical		0.8	0.5		μW	*
Receiver	Sensitivity		−31	−33		dBm	
HFBR-3000 Series Cable/Connector	Insertion Loss	Length Dependent		7	10	dB/km	*λ = 820nm ℓ > 300m
		Fixed		5.4	8.4	dB	*λ = 820nm ℓ ≤ 300m

*NOTE: Guaranteed specifications 0°C–70°C, ±5% Voltage, 10^{-9} BER @ 10 Mbaud.

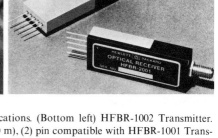

Figure 4.19. (Top) Data bus systems specifications. (Bottom left) HFBR-1002 Transmitter. Features: (1) long distance transmission (1000 m), (2) pin compatible with HFBR-1001 Transmitter, (3) high speed (dc to 10 Mbaud), (4) no data encoding required, (5) functional link monitoring, (6) TTL input levels, (7) built-in optical connector, (8) low profile: PCB mountable, (9) single +5 V supply. Note features 1, 3, 4, and 5 apply when used with HFBR-2001 Receiver Module and any Hewlett–Packard HFBR-3000 Series Cable/Connector Assembly. (Bottom right) HFBR-2001 Receiver. Features: (1) high speed (dc to 10 Mb/s NRZ), (2) low noise (10^{-9} BER with 0.8 μW input), (3) low profile [fits 12.7 mm (0.5 in.) spaced card rack], (4) single supply voltage, (5) wide optical dynamic range (23 dB), (6) optical port connector, (7) arbitrary data format, (8) TTL output levels, (9) link monitor (shows satisfactory input signal). Note: features 1, 2, 7, and 9 apply when used with HFBR-1001 Transmitter and HFBR-3001 to −3005 Cable/Connector Assemblies. (From Hewlett–Packard published data.)

"regenerative tap" approach, shown in Figure 4.22, the optical signal terminates on the terminal and a new optical signal is generated by the terminal. Digital information can be modified in the terminal as it passes through (bits are added and removed). An advantage to this method is the simplicity in the requirements on optical components. Only sources and detectors are required, just as in the point-to-point link. However, a dis-

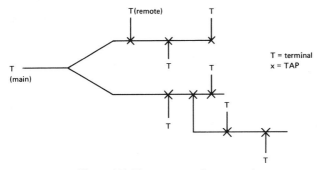

Figure 4.20. Tree structure data network.

advantage to this approach is that a failure of the electronics in a single terminal can fail all or most of the network. In the "optical tap" approach, shown in Figure 4.23, an optical coupler is used to remove light from a fiber or to add light to a fiber. The physical realization of the optical tap can be one of the devices shown in Figure 4.24.

If the terminal electronics fail, only the particular affected terminal loses communication. Disadvantages of this method are the need for an optical tap and the inability to completely remove or to modify pulses on the optical highway. A major problem with the optical tap involves the optical power budget of the network. Since the optical signal is not re-generated at each terminal node, tap losses (insertion, input, and output) and fiber losses between taps seriously limit the network size.

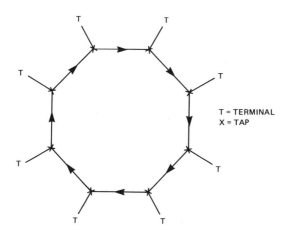

Figure 4.21. Ring structure data network.

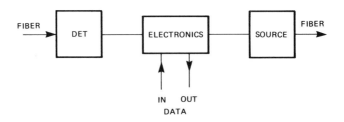

Figure 4.22. Regenerative tap terminal.

An approach which has the advantages of the regenerative tap, but which is more fault tolerant, is shown in Figure 4.25. Here an optical switch is used to shunt a failed terminal. The obvious disadvantage here is the need for the optical switch and its associated cost, insertion loss, and reliability.

4.2.3. Long-Haul Systems

Long-haul systems are similar to the trunks described above in Section 4.2.1 except the transmission distance can be hundreds or thousands of miles rather than a few miles or a few tens of miles. Because of the long transmission distances, it is economical to gather many circuits together ("back haul-to-backbone" approach) for multiplexing to a very high data rate. Data rates over 100 Mbaud (corresponding to over a thousand multiplexed telephone channels) are typical. Because of the desire to maximize the distance between outside plant repeaters (see Figure 4.26) one would in general use laser sources to obtain high input powers and low chromatic dispersion. One would use low-loss optical fibers with large bandwidths (low delay distortion) either of the graded index type or, if practical, single-

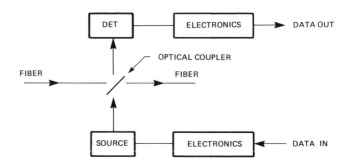

Figure 4.23. Optical tap terminal.

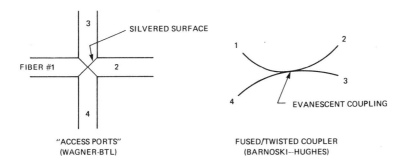

Figure 4.24. Examples of couplers.

mode type. To minimize the fiber loss, one would be tempted to operate, if practical, at longer wavelengths (near 1.3–1.5 μm) where Rayleigh scattering is lower. Thus one would look closely at the newer InGaAsP sources rather than the older and more fully developed GaAs sources.

In long-haul systems, repeater powering and ambient control is more of a problem than in short-haul systems (where repeaters could be located in ambient-controlled buildings with local power). Repeaters can be powered over copper wires in the fiber cable or over an auxiliary copper pair cable. This is somewhat undesirable because metal in the cable makes the fiber system more susceptible to lightning damage (mechanical and electrical). Alternatively, local ac power with battery backup could be used at repeater sites. This is a problem because of the cost and the difficulty of maintaining a battery plant at remote location. Ambient control is a very serious problem since the lifetime of a typical GaAs source decreases by a factor of 10 for a 20°C increase in device-operating temperature. In outside-plant huts or housings the maximum ambient temperature can exceed 80°C,

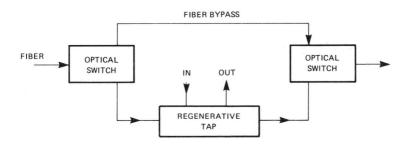

Figure 4.25. Fault-tolerant tap.

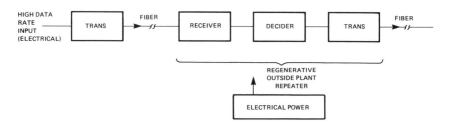

Figure 4.26. Long-haul optical trunk.

whereas in buildings, ambient temperatures are usually expected to remain below 50°C, even when air conditioning systems fail.

Perhaps the most challenging long-haul fiber application is the trans-oceanic undersea cable. Here the highest priority is placed on reliability. A typical analog coaxial transoceanic undersea cable might have repeaters spaced every 10 miles for a total of 500 repeaters. Each repeater might contain about five active devices. The desired mean time between failures for the system is about 20 years, with a failure requiring retrieval of the submerged cable and repeater. With a total of 2500 active devices in the 5000-mile span we require each device to have a nominal mean time between failures of 50,000 years or a failure rate of about 2 FITs (failures in a billion device hours). Each repeater must have a failure rate of about 10 FITs. Reliabilities of this kind are obtained in existing transoceanic systems (at great expense) by extreme care in the fabrication and in the testing of the active devices. If a fiber optic digital repeater for this application were to be developed, one would have to fabricate high-speed integrated circuits for the electronics, optical sources, and optical detectors with 1–10-FITs reliability.

4.3. Examples of Analog Systems [3, 11, 13]

There are several situations where a signal to be transmitted from one point to another cannot conveniently and economically be converted to digital form. One example in telephony is an FDM (frequency-division-multiplexed) "mastergroup" consisting of 600 4-kHz voice channels, stacked in frequency from 60 to 2788 kHz. In principle such a signal could be sampled and encoded into a binary signal, but the linearity and fidelity requirements on the coding and subsequent decoding in order to prevent "crosstalk" between the channels is severe. An alternative would be to demultiplex the FDM signal to individual analog voice channels and to recombine them in digital form using digital channel banks and a time division multiplexer. This back-to-back demultiplexing–multiplexing is expensive. Thus it would be preferable to build a fiber optic system which could transmit this signal in analog form.

Another situation where analog transmission is preferable involves video signals. A typical base-band video signal occupies about 5 MHz of bandwidth. Coders exist to convert such signals to digital form. Simple coders sample the signal at about a 10-MHz rate, and code the samples to nine binary bits each. Thus a simple coder requires 90 Mbaud (binary) of digital transmission capacity per video channel. More sophisticated coders take advantage of the redundancy of most video signals. They use large signal storage capacities to transmit only those portions of the picture which are changing. Adequate picture quality can be obtained with such sophisticated terminals with less than 10 Mbaud. However, at present the cost of these terminals is so large (around $25,000–100,000 per end) that they are practical only when used on very long transmission links (e.g., over digital satellites).

CATV systems transmit several video signals in analog FDM format so that they can easily interface to standard home receivers. Here again the cost of digital coders is at present prohibitive, except perhaps on the "supertrunk" portions of the CATV distribution network.

There are several methods of modulating an optical source with an analog signal. These include intensity modulation.

4.3.1. Direct Intensity Modulation

The simplest analog modulation scheme for fiber optic systems is direct intensity modulation as shown in Figure 4.27. A message, $m(t)$, is used to modulate the source output power $p_s(t)$ above and below a bias value as follows:

$$p_s(t) = P_{os}(1 + k_1 m(t)) \qquad (4.3.1)$$

where $m(t)$ is adjusted to have unity peak value and is assumed to have zero average value, k_1 is a modulation index less than unity, and P_{os} is the average source output power (watts). Such modulation can be accomplished

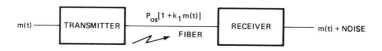

Figure 4.27. Simple intensity modulation system.

with an LED or a laser by varying the current driving the source, above and below a bias point.

At the receiver the incoming power, which is an attenuated and band-limited version of $p_s(t)$, is converted to a current, amplified, and filtered (to limit the noise).

Neglecting distortion for the moment, the receiver output waveform is of the following form:

$$v_{out}(t) = m(t) + n_s(t) + n_{th}(t) + n_q(t) \qquad (4.3.2)$$

where $n_s(t)$ is noise associated with fluctuations in the source (e.g., noise caused by mode hopping in a laser), $n_{th}(t)$ is thermal noise introduced in the receiver, and $n_q(t)$ is quantum noise in the detection process enhanced perhaps by the randomness of avalanche multiplication.

Signal-to-noise ratios, at the receiver output, for direct-intensity-modulated analog systems were discussed in Section 3.4.5 above, except for source fluctuation noise, which was discussed briefly in Section 4.1.2 above. For simplicity here, let us consider quantum-noise-limited receivers where the peak signal-to-rms-noise ratio at the receiver output is given by

$$\text{SNR}_{q.l.} = \frac{k_1^2 P_{or}}{2hfB} \qquad (4.3.3)$$

where B is the message bandwidth (hertz), hf = energy in a photon $\cong$ 2×10^{-19} (joules), and P_{or} is the received average optical power (watts).

Consider now a system with an LED transmitter emitting an average power of 25 μW into a fiber. Assume that the modulation index k_1 is 0.5. Let the message $m(t)$ consist of a single video signal with a required bandwidth of 5 MHz. Let the required signal-to-noise ratio at the receiver output (peak signal to rms noise) be 50 dB. Then from (4.3.3) above we require a minimum received average power P_{or} of

$$P_{or} = (10^5) \times 2 \times (2 \times 10^{-19}) \times (5 \times 10^6) \times (2)^2 = 0.8 \times 10^{-6} \text{W} \quad (4.3.4)$$

Thus even under ideal quantum-limited conditions with this source and with this required SNR, the allowable attenuation between the source and the receiver is only a factor of 31.2 or 15 dB. Optimistically one could assume that for a 5-dB/km fiber loss this implies up to 3 km of fiber spacing. However, when one takes into account the receiver thermal noise (which we have neglected above) and if one allows for connector losses and component variations, one finds that even a very short link of this sort has very little margin for error.

One saving feature of such an analog link is that the performance degrades "gracefully" as the received signal level drops. If the signal level is 10 dB low, then the SNR is 10 dB low, which may or may not be noticeable to the user.

The situation worsens when one wishes to transmit several frequency-division-multiplexed video channels. The required bandwidth increases in proportion to the number of channels, and from (4.3.3) so does the required optical power at the receiver. The use of a laser source can help the situation by increasing the transmitted average power by 10–20 dB. However, laser sources suffer from source fluctuation noise, and also may not have the necessary linearity to meet video FDM crosstalk and differential phase and gain requirements.

It is interesting to note what happens when analog fiber links are placed in tandem. Consider the example above, where the allowed loss between the transmitter and the receiver was 15 dB (optimistically). Suppose one places N such fiber links in tandem, but requires that the signal-to-noise ratio at the output of the last receiver in the chain still be 50 dB. Then, since the noises in an analog system add (in power), each link must have a signal-to-noise ratio at its receiver output of $50 + 10 \log N$ (dB). This implies that the loss between each transmitter and receiver must decrease by $10 \log N$ (dB). Figure 4.28 shows how the allowed loss between the transmitter and receiver on each link varies with N, and how the total allowed loss for the overall chain of repeaters varies with N. We see that with an end-to-end signal-to-noise ratio requirement, there is a point of diminishing returns ($N = 12$ in this case) where placing more links in tandem does not increase the overall transmission distance (50.5 dB of loss).

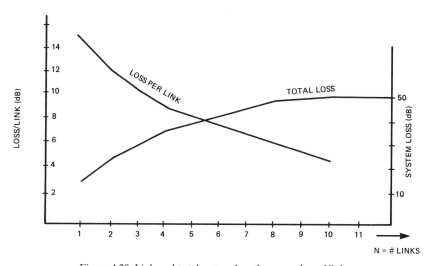

Figure 4.28. Link and total system length vs. number of links.

4.3.2. Comparison of Fiber and Metallic Cable Systems

In order to understand why fiber systems are in general more suited to digital modulation than analog modulation, it is interesting to compare with metallic cable systems where the reverse is true. Consider the signal-to-noise ratio at the transmitter for each of these systems. This is defined as the signal-to-noise ratio at the output of a receiver connected directly to the transmitter (with no intervening attenuation and, of course, neglecting overload):

$$\text{SNR}_{\text{fiber, quantum limited, at the transmitter}} = P_{os}k_1^2/2hfB \qquad (4.3.5)$$

$$\text{SNR}_{\text{metallic, thermal limited, at the transmitter}} = V_s^2/4kTBZ_0$$

where P_{os} is the average output power of the fiber system transmitter (watts), k_1 is the intensity modulation index for the fiber system, V_s is the peak output voltage for the metallic system transmitter (volts), $kT = 4 \times 10^{-21}$ (watts/hertz), Z_0 is the metallic cable impedance (ohms), and B is the bandwidth of the message (hertz).

We see that for the fiber system if we set $P_{os} = 25$ μW (for an LED) and $k_1 = 0.5$, we obtain

$$\text{SNR}_{\text{fiber, quantum limited, at the transmitter}} = \frac{1.5 \times 10^{13}}{B} \qquad (4.3.6)$$

For the metallic cable system, if we set $V_s = 3$ V, and $Z_0 = 100\,\Omega$, we obtain

$$\text{SNR}_{\text{metallic, thermal limited, at the transmitter}} = \frac{5.6 \times 10^{18}}{B} \qquad (4.3.7)$$

Thus we see that, at the transmitter, a typical metallic cable system has about 55 dB more signal-to-noise ratio than an LED fiber system. With a laser, the difference would still be about 40 dB.

On the other hand, for metallic cables, the loss is an increasing function of frequency, owing to skin effect. For fiber systems, the attenuation is independent of the base band modulation frequency up to the point where delay distortion becomes important. Figure 4.29 shows losses per kilometer vs. base-band frequency for typical metallic and fiber cables.

The net result of this is that metallic systems perform best in an analog mode with a high signal-to-noise ratio, but with a low-bandwidth requirement. Fiber systems perform best when signal-to-noise ratio is traded against available bandwidth using digital modulation or other bandwidth expansion techniques.

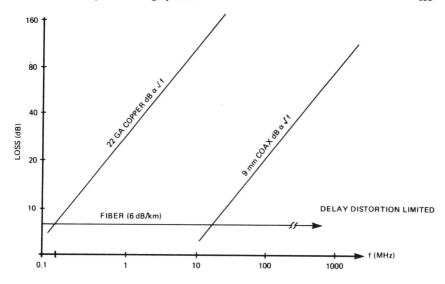

Figure 4.29. Loss/km vs. baseband frequency for typical fiber plus metallic cables.

4.3.3. Analog Modulation Schemes Which Expand Bandwidth

As derived above, fiber optic systems work best when available bandwidth is traded for precious signal-to-noise ratio at the receiver output. Digital modulation does this by expanding bandwidth by a factor of roughly 15–20. Thus, for example, a 4-kHz voice channel is coded by a standard "channel bank" to a 64-kbaud binary data stream. The signal-to-noise ratio required at the output of a digital receiver is about 20 dB (for a negligible error rate). The equivalent signal-to-noise ratio at the output of the channel bank (after decoding back to analog) is about 70 dB (helped by what is called "companding" in the analog sample-to-digital coder).

When digital encoding is not practical (e.g., for economic reasons) there are various analog techniques for bandwidth expansion.

Subcarrier frequency modulation[11] (also called subcarrier FM–IM, for "subcarrier FM intensity modulation") is popular for several reasons. In this method, the message $m(t)$ frequency-modulates an "intermediate frequency" carrier (e.g., at 70 MHz), which in turn intensity-modulates the optical power, as shown in Figure 4.30.

$$p_s(t) = P_{os}\left[1 + k_1 \cos\left(\int_{-\infty}^{t} 2\pi d_f m(u)\, du + \omega_{\text{i.f.}} t \right) \right] \qquad (4.3.8)$$

where P_{os} is the average transmitter output power (watts), k_1 = intensity

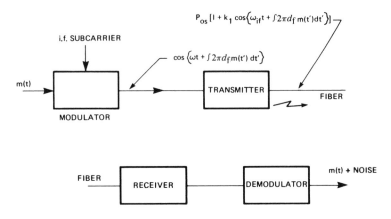

Figure 4.30. FM subcarrier intensity modulation.

modulation index ≤ 1, $\omega_{\text{i.f.}}$ is the intermediate frequency of the subcarrier rad/s, d_f is the peak frequency deviation, and m(t) is the message with bandwidth B and unity peak value.

One reason for the popularity of subcarrier FM is that many conventional microwave radio systems use frequency modulation to transmit frequency-division-multiplexed (FDM) voice signals and video signals. Thus the equipment to generate the FM intermediate frequency carrier may be available. Similarly, a fiber optic system which can interface directly with such an i.f. carrier can interface economically with a microwave radio system which also uses this modulation format.

Because the FM subcarrier is not very sensitive to distortion, one can use a large modulation index, k_1, without being concerned with linearity. This also makes laser transmitters more applicable than in the intensity modulation approach.

FM systems can expand the bandwidth required (with a corresponding reduction in i.f. signal-to-noise ratio requirements). For the modulation given in (5.3.8) the bandwidth occupied by the i.f. subcarrier is roughly $2d_f$. The bandwidth expansion is therefore $2d_f/B$, where B is the base-band bandwidth of the message $m(t)$. Classical modulation theory shows that for FM the signal-to-noise ratio required at i.f. is less than the signal-to-noise ratio required at base band by approximately the factor $6(d_f/B)^3$. Thus if, for example, d_f is 9 MHz and B is 3 MHz, then the signal-to-noise ratio required at i.f. is reduced by 22 dB relative to base band. Since in this example the i.f. bandwidth for FM–IM is increased over direct intensity modulation by a factor of 18/3, the net reduction in the required optical signal at the receiver input, relative to direct intensity modulation, is 14 dB, assuming quantum limited operation. By combining the reduction in the required received optical power associated with bandwidth expansion, with the increase in the modulation index and the increase transmitter output (using lasers instead of LEDs) allowed by the relaxed linearity requirements, net increases in the allowed loss between the transmitter and the receiver of 20 dB or more are possible, using subcarrier FM − IM. However, even with the relaxed signal-to-noise ratio requirements at i.f., noise at the receiver associated with laser output fluctuations (due to the mode hopping and other causes) is still a serious problem.

Another bandwidth expansion technique often used with laser systems is pulse position modulation. Here the message must be band limited and sampled periodically as shown in Figure 4.31. The sampling rate must be twice the bandwidth of the band-limiting filter to avoid "aliasing." The optical transmitter emits a pulse stream as shown, where in each time slot the position of the transmitted optical pulse relative to the center is controlled by the corresponding analog "sample" amplitude. At the receiver output, the analog samples are reconstructed by estimating the positions of the received pulses relative to the centers of their time slots using a threshold-crossing detector. The errors in the reconstruction of the analog samples are controlled by the rise times of the receiver output pulses, and by the

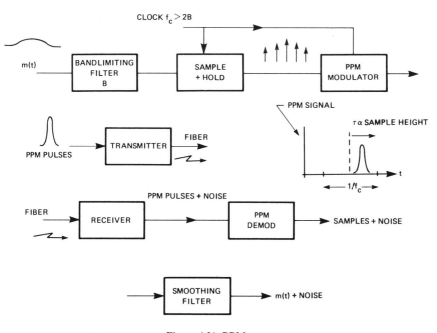

Figure 4.31. PPM system.

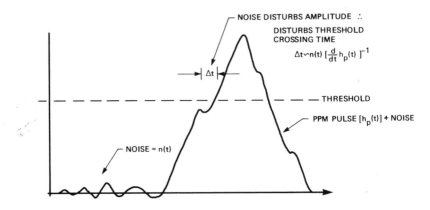

Figure 4.32. Local PPM errors due to noise.

noise level at the receiver output. Figure 4.32 shows how noise at the receiver output can result in an error in the time of a threshold crossing of a
receiver output pulse.

There are two types of errors which can occur at the receiver output.
One is a "local error" as shown in Figure 4.32 where, as mentioned, noise
slightly disturbs the time of the threshold crossing. "Global errors" are
shown in Figure 4.33. Here noise causes a false threshold crossing, or completely eliminates a desired threshold crossing. Figure 4.34 shows the quali-
tative performance of a pulse position modulation system as a function of
the received optical power level and the bandwidth of the combined transmitter, fiber, and receiver (inverse of receiver output pulse width). We
observe that for a fixed received power level the base-band SNR in the
local error region increases with increasing system bandwidth (i.e., narrower
transmitted pulses and more receiver bandwidth). However, as the system

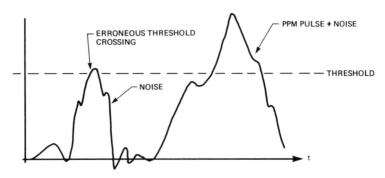

Figure 4.33. Global errors due to noise.

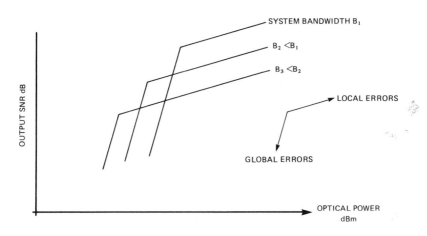

Figure 4.34. PPM performance.

bandwidth increases, the probability of a global error increases. Thus the received power level threshold for global errors increases as the bandwidth increases. The exact performance of a pulse position modulation system depends on many details such as whether a PIN detector or avalanche detector is used, whether the transmitter is peak power limited or average power limited, whether the fiber bandwidth is a limiting factor, etc. The advantage of such systems is the tradeoff of bandwidth for required signal-to-noise ratio at the threshold-crossing detector. Such systems are also ideally suited for use with certain optical sources, which work best when they emit a low-duty-cycle sequence of narrow, high-energy optical pulses.

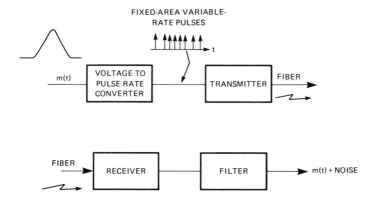

Figure 4.35. Pulse frequency modulation.

The linearity at base band is controlled by the accuracy of the pulse amplitude-to-pulse position converters at the transmitter and at the receiver.

Another analog modulation scheme which does not require sampling of the message is shown in Figure 4.35. The amplitude of the message controls the pulse repetition rate of an optical transmitter which emits pulses of a fixed energy. Thus the message modulates the average power emitted by the source. The receiver is the same as for direct intensity modulation. The requirements on fiber bandwidth are also the same as for direct intensity modulation. The advantage of this system is that the linearity at the transmitter is controlled by the fidelity with which the pulse repetition frequency tracks the message amplitude. Here again laser sources can be used provided laser noise (in this case manifesting itself as variations in the energies of the received pulses) does not cause problems.

Problems

1. A source emits 1 mW and has a spectral width of 50 nm. It has a circular emitting area 100 μm in diameter and a half angle of emission 45°. What is its brightness in W cm^{-2} sr^{-1}? It emits at 0.8-μm wavelength. How many modes does it emit into? What is its energy/space-time mode?

2. A fiber has an index of refraction of 1.5, a core diameter of 50 μm, and an N.A. of 0.2. How many modes does it propagate at 0.8 μm wavelength?

3. What coupling efficiency can be attained between the source and fiber described above?

4. A source emitting at 0.85 μm wavelength has an rms spectral width of 50 nm. A fiber has a dispersion of 100 ps km^{-1} nm^{-1} and an rms model delay spread of 2 ns km^{-1}. What is the total rms impulse response width per kilometer when the source is used with this fiber?

5. A 0.82-μm wavelength laser emits 1 mW average power into a fiber. A corresponding receiver requires 500 photons per received pulse for a 10^{-9} error rate. Calculate and plot the allowed transmission distance in kilometers, for 4 dB km^{-1} fiber loss, between the transmitter and the receiver vs. bit rate for 0.1–1000 Mb s^{-1} (allow 10 dB of margin). Assume the fiber delay distortion is 2 ns km^{-1} rms and that the allowed rms delay distortion must not exceed 0.25 time slots. Calculate again the allowed transmission distance for 0.1–1000 Mb s^{-1}.

6. Repeat Problem 5 for an LED source emitting 0.05 mW into a fiber. Assume that the LED has spectral width of 50 nm and that the fiber dispersion is 100 ps nm^{-1} km^{-1}

7. Repeat Problem 5 assuming that the fiber model delay spread is 2 ns km^{-1} for lengths up to 1 km, and increases in proportion to the square root of length thereafter.

Applications [1-4]

Having discussed the technology of optical fiber systems, subsystems, and components in the previous four chapters, we shall now discuss the applications for these systems and system pieces. Roughly speaking, these applications can be divided into two categories: price driven and feature driven. The distinction is not black or white, but the differences in emphasis are important to understanding how and when fiber systems will be used.

Price-driven applications are characterized by a competition of technologies. That is, there will be a number of alternate ways to provide the desired service or function, many of which are perfectly satisfactory to the customer. Each technology has some advantages and disadvantages over the others, but no one advantage is so critical to the application that a given technology becomes an obvious best choice. Often, but not always, there will be some "classical" technology in widespread use. To displace this classical technology, a new alternative must show a large projected cost savings or substantial new features (in order to overcome the legitimate inertia toward staying with proven methods). If the price (and maintenance cost) advantage is not overwhelming, then the new technology will require considerable amounts of "selling" and numerous "trial" applications before it can displace the old. These early applications will uncover unforeseen advantages as well as disadvantages with the new approach. As confidence builds in projected costs, and as customers become familiar with the new technology, it will replace the old with an accelerating pace of introduction (provided it is indeed proven to be more cost effective).

Feature-driven applications are those where a particular attribute of a new technology is so valuable that there is no question that the new approach will be cost effective. In some cases, the new feature or features make possible services or functions that were not possible before. The introduction pace of the new technology in these applications is limited by availability, and not by customer inertia.

5.1. Telephone Trunk Applications

As mentioned in Chapter 4, telephone "trunks" are the voice-carrying circuits that go between telephone buildings. These buildings may be less than a mile apart (e.g., some telephone offices in Manhattan) or thousands of miles apart. As we shall see below, the preferred technologies in present usage for providing these circuits are different for different distances of transmission. Telephone trunks were the first major applications considered for optical fiber systems. One reason for this is the huge size of the potential cost savings that a more economical technology could bring. The present investment in trunk transmission facilities is several tens of billions of dollars in the United States alone. The annual U.S. expenditures for trunk circuit growth and replacements is several billion dollars. In order to understand how fibers can find applications in the trunk "plant" it is useful to consider the existing approaches to providing trunk circuits in some detail.

5.1.1. Metallic Trunk Facilities

The simplest way to provide a voice circuit between two telephone buildings is to use two pairs of wires, one for each direction. Typically these wires are contained in cables of varying sizes. In metropolitan applications, where large circuit cross sections are required, cables containing 1000–2000 pairs of wire (about 3 in. in diameter) are pulled through concrete ducts installed beneath the streets. In suburban and rural areas, smaller cables containing several tens or hundreds of pairs are buried in trenches or installed on poles. The cost per mile per circuit (two pairs required) depends not only on the cost of the cable, but also very much on the installation costs. As an extreme example, a 3-in.-diameter, 900-pair cable for metropolitan use might cost $25,000 per mile (material). The duct it occupies might cost more than a million dollars a mile to install in downtown Manhattan. The cost of installing the cable in the duct and splicing together the installed sections may also exceed the cost of the cable material. For the sake of this comparison of technologies we shall use a nomimal cost per mile of two pairs in an installed cable of $100. Trunk circuits which are provided on individual wire pairs are called "metallic trunks."

5.1.2. Paired Cable Carrier Systems

To reduce the cost of providing trunk circuits between buildings one can combine several circuits together at the ends of the transmission system using a time division multiplexer (TDM) or a frequency division multiplexer (FDM). The multiplexed signals can then be transmitted over wire pairs in a cable, with the wire pair costs shared amongst the multiplexed circuits. A system which uses this approach is called a "carrier" system. If the savings in transmission costs is larger than the cost of the multiplexers, then the carrier system will "prove in." The cost of an analog channel bank which combines (multiplexes) several tens of analog circuits in an FDM fashion is roughly $500 per voice circuit (two way) at each end of the transmission system. Thus the terminal cost "two way both ends" (TWBE) is about $1000 per voice circuit. A standard (in North America) digital channel bank which combines 24 voice circuits into a pulse code modulated (PCM) format at a 1.544–Mbaud rate costs about $250 per voice circuit (TWBE). The digital carrier system requires more bandwidth per voice circuit (about 32 kHz per voice circuit for digital vs. about 4 kHz per voice circuit for analog). Thus, with the frequency-dependent loss and crosstalk of cable pairs, the number of voice circuits which can share a pair is larger for analog than for digital carrier. Figure 5.1 is a field-of-use chart showing a cost comparison (dollars per voice circuit mile) of metallic,

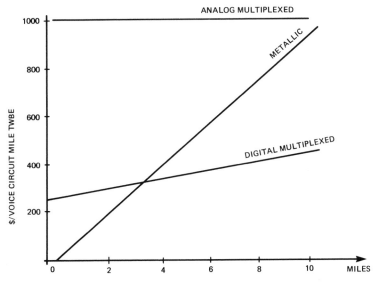

Figure 5.1. field of use chart.

analog carrier, and digital carrier systems. A nominal transmission cost of $20 per mile (including repeaters) per voice circuit is shown for the digital carrier system. The cost per voice circuit per mile for the analog system is nominally shown as $5.00. We see that the digital carrier system "proves-in" over the metallic system at a distance of 3 miles (nominally). The analog carrier system does not prove-in over digital until an extrapolated transmission distance of 50 miles. (There are several technical issues beyond the scope of this book which limit the actual distance one can transmit using analog carrier on wire pairs, particularly at high circuit capacities per pair.)

5.1.3. Coaxial Cable Carrier Systems

Multiplexed carrier systems can be built using coaxial cables rather than wire pairs for the transmission medium. The larger cross section and reduced crosstalk of coaxial cables allows significantly increased numbers of voice circuits to share a pair of coaxial tubes (vs. two-wire pairs). Analog (FDM) carrier systems for use with 0.375-in.-diameter coaxial tubes have been built with more than 13,000 voice circuits per tube (more than 130,000 two-way voice circuits in a 3-in.-diameter 20-tube cable). Digital (TDM) coaxial systems have been built with more than 4000 voice circuits per tube. The analog coaxial systems are used for long-distance trunk routes with sufficient cross section (circuit requirements) to fill up the system. Digital coaxial systems are used in metropolitan applications where very large cross sections and moderate transmission distances are required. It is important to note that the additional multiplexing costs of combining the outputs of 168 digital channel banks (1.544-Mbaud signals containing 24 voice circuits each) to produce a high-speed digital signal (274 Mbaud, 4032 voice circuits) must be offset by a savings in transmission cost, before the high-speed multiplexed digital coaxial system can prove-in over the wire pair digital system. In metropolitan applications the requirements for large cross sections and moderately long transmission distances (5–10 miles) are somewhat contradictory. The large cross sections tend to be required between nearby locations with a large community of interest. Nevertheless, high-capacity digital coaxial systems do find their applications, particularly when network rearrangements are made to take advantage of these high-capacity systems.

5.1.4. Microwave Radio Systems

Trunk circuits can be transmitted from place to place using "line-of sight" microwave radio systems. The advantage of radio is the cost savings from not having to install and maintain cable plant and "outside plant" repeaters (including the purchase of "right of way" and installation costs).

However, radio systems can only be installed where there are sufficient microwave frequency allocations available, where terrain permits line-of-sight installations, and where the savings in transmission are sufficient to offset the costs of radio equipment, towers, antennas, etc. Microwave radio systems are popular for long-distance trunk routes in sparsely populated areas. They are also useful in areas where geological conditions make cable installations difficult (e.g., rocky ground, crossing the Grand Canyon, etc.). In some sense, the use of microwave systems rather than cable is often feature driven.

5.1.5. Proving-in Fiber Systems

There are both feature-driven and price-driven applications for optical fibers in the telephone trunk plant. On the feature-driven side, metal-free fiber cable systems offer significant advantages over copper cable systems in areas where lightning or induction are severe (e.g., telephone circuits near power-generating stations or power transmission cables). By the same token, nonmetallic fiber cables do not present a spark hazard in explosive environments (e.g., telephone circuits on the premises of a chemical plant). Fiber cables may survive in environments which are corrosive to copper cables. In metropolitan areas the small size of fiber cables can save precious underground duct capacity. For example, a 900-pair 3-in.-diameter copper pair cable has a capacity of about 10,000–20,000 two-way voice circuits. Fiber cables less than 1 in. in diameter with a capacity of 50,000 two-way voice circuits have been demonstrated in the field. In addition to saving space, fiber cables of this small size can be installed in long continuous lengths (perhaps as long as a mile). Compared to 1-mile repeater spacings with wire pair cables, fiber cable systems offer repeater spacings of 4 miles or more. (In the laboratory repeater spacings of more than 30 miles have been demonstrated.) These long repeater spacings can help eliminate "outside plant" repeaters, located in manholes or on telephone poles. Such outside plant electronics can be very expensive (or impossible) to install and maintain under some circumstances.

On the price-driven side, the prove-in of fiber cables is analogous to the prove-in of digital coaxial trunk carrier systems. The easiest way to apply fiber cables to the trunk circuit plant would be to replace wire pair digital carrier in metropolitan and other applications. In the near term (5 years) fibers are expected to be considerably more expensive than wire pairs; even in high volume production. Projections are for fiber cable (material) prices to drop to about $500–$1000 per fiber mile compared to about $25–$50 per wire pair mile. To be economically competitive the number of circuits transmitted per fiber will generally have to be 10–100 times larger than the number of circuits transmitted per wire pair. This presents two problems. The transmission distances must be long enough for the transmission savings with fibers to offset the cost of the multiplexers required to combine these circuits together; and the required circuit cross section must be adequate to fill up the fiber systems. For metropolitan applications much attention has been focused on the 672 voice circuit, 44.7 Mbaud, rate (28 times the capacity of a 1.544 Mbaud, 24 voice circuit, system). For rural and suburban digital trunk carrier applications this rate may be too large to obtain the required fill.

For long-distance trunk applications projected costs of high-capacity (more than 100 Mbaud) digital fiber systems look very promising compared to analog coaxial cable systems. Thus for new installations the choice will probably be between radio and fiber systems, on a case-by-case basis depending upon whether the features of radio (no cable installation) outweigh the disadvantages (line of sight, expensive terminals).

For growth on existing long-distance trunk routes there is an additional factor. Many existing facilities can be upgraded to higher circuit capacities by retrofitting the electronic repeaters (on cable systems) or the terminal equipment (on radio systems) with improved or additional componentry. The cost of such "retrofits" and "overbuilds" is generally much lower than the cost of installing a new cable, fiber or otherwise. Thus on existing long-distance routes, the introduction of fiber cables may be delayed until such upgrade capabilities are exhausted.

An interesting footnote to the subject of fiber optic telephone trunk applications is that in an actual system a great deal of the cost is in the electronic terminal equipment (multiplexers) and in the electronics in the repeaters. The amount of the total cost in the critical optical sources and detectors (assuming present projected prices) is relatively small. Even the cost of the fiber cable may be less than 1/3 of the total materials cost for the system. This is analogous to the new digital telephone switching machines, where the amount of materials cost in the critical microprocessor and peripheral LSI devices is a very small part of the total system cost.

5.2. Telephone Loop Applications

Telephone "loops" (or "lines") are voice circuits which go between a telephone building and a telephone customer. Unlike trunks, which are shared on a need-to-use basis, loops are in use only when a specific customer needs service (except for party lines, which are shared amongst two, four or eight customers). Most loops are provided using twisted copper wire pairs, with a single pair used for both directions of transmission. The transmit and receive directions at each end are separated using devices called "hybrids." In addition to the obvious function of providing a voice channel with adequate bandwidth and signal level, the loop must perform several other important functions called "BORSHT" functions. These include: the application of a low-frequency high-voltage ringing signal at the telephone office end to ring the customer's phone; application of a direct current to power the customer's phone and to detect the current which flows when the customer is "off-hook"; providing a means of testing the condition of the loop and the phone at the other end, and transmitting dial pulses to the telephone office from the customer's telephone set. Any transmission system that is proposed to replace the wire pairs must be capable of providing these BORSHT functions one way or another.

One method for doing this is applicable to long loops (more than a few miles). When customers are too far from a telephone office, the cost of providing a pair of wires per customer is often prohibitive. Further, the quality of the service that can be provided over a long loop is poor. To alleviate these problems, one can use a "subscriber carrier system" as shown in Figure 5.2. A remote (unattended) electronic subscriber terminal is placed near a group of distant customers. This terminal is powered by local ac mains, with battery backup in the event of a power failure. Customers' loops terminate on this remote terminal. The terminal can ring phones, provide dc power to the loops, etc. The required signals are generated locally within the terminal. Commands from the office to the terminal to execute BORSHT functions (e.g., ring a particular phone) and inputs from the terminal to the office (e.g., customer A has gone off-hook) are communicated over a carrier transmission system. The carrier transmission system also provides the voice circuits (in FDM or TDM multiplexed

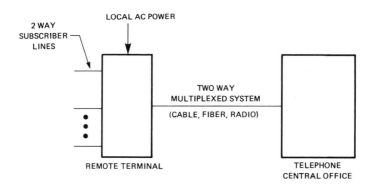

Figure 5.2. Subscriber carrier system.

format). Thus many customers can share a few pairs of wire or a radio channel or whatever carrier transmission system is used. The quality of service provided to the remote customers is governed by their distance from the subscriber terminal (and the quality of the terminal and carrier system) rather than by the distance of the subscribers from the telephone office.

To prove-in a fiber optic transmission system in this application is analogous to proving-in optical fibers in the short-to-medium distance trunk application described above. The main competitive transmission technologies (presently used) are time division multiplexed carrier using wire pairs, frequency division multiplexed carrier using wire pairs and microwave radio systems. The feature-driven applications are motivated by the same considerations as for trunks: a non metallic cable, more circuit capacity per unit cable cross section, and fewer or no outside plant repeaters.

The price-driven applications are those where the required circuit cross section is sufficient to justify the use of a high-speed multiplexed fiber system. In this regard it is interesting to note that there are types of remote terminals that terminate hundreds of subscribers. These are usually called remote "switches" or "concentrators" because they provide fewer available circuits back to the telephone office than the number of subscribers they serve. (Customers can receive service only if an unused circuit is available.) The concentrating function reduces the number of copper pairs required in the transmission link between the remote terminal and the office. By using fiber optic transmission, large numbers of remote customers might be served without this concentration function or with less concentration. Whether this is advantageous economically has not been resolved. Alternatively, fiber transmission systems may prove-in over wire pair cable systems when even larger numbers of remote customers are served by even larger remote terminals. One should point out that a connection between a remote switch (concentrator) and the telephone office is really a trunk, rather than a loop. Thus if fiber optic transmission somehow motivates larger remote switches, this is really a change in the configuration of the trunk network.

An alternative to the use of subscriber carrier systems and remote switches is to replace the customer's loop directly with a fiber. The challenge here is very severe since not only must the fiber be more economical than the wire pair it replaces, but it must also somehow provide the BORSHT functions. This implies, amongst other things, a more complicated telephone or a terminal at the customer's premises. One motivation for providing a fiber loop (perhaps in addition to rather than in lieu of a wire loop) is to provide the capacity for video transmission to and from the customer's premises. It is not yet resolved whether this service is best provided via fibers or by other means (as will be discussed in Section 5.3).

5.3. Video Distribution Applications

Figure 5.3 shows a typical present-day entertainment video distribution network. Cameras and video playback equipment in studios generate video signals of a given quality. These are transmitted to broadcast stations over video loops as shown. The signals from the broadcast stations can be carried to customer receivers directly or via cable television facilities. Cable television facilities intercept the signals from several broadcast stations and combine them in frequency division multiplexed form for transmission over coaxial cables. With rare exceptions, all video signals are transmitted from place to place in analog form. In the process of being transmitted from the camera or playback equipment to the customer's receiver, various degradations are introduced in the video signals. These include band limiting, distortion, noise, and interference from other signals. Because the signals are transmitted in analog form, degradations from various portions of the transmission network accumulate. To guarantee a reasonable quality of reception, an allowable total (end-to-end) transmission degradation must be established, and allocated amongst the various portions of the network. When doing this allocation one must take into account worst-case conditions, including the maximum number of links which can exist in the path to the customer from the video source; margin for worst-case link performance; and realistic margin for improper "network engineering"

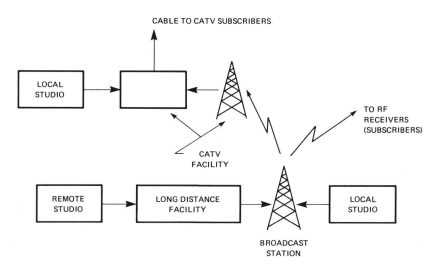

Figure 5.3. Entertainment video distribution network.

or installation of the equipment. This worst-case analysis approach assures that the network will be viable, and in particular maintainable with a reasonable incidence of trouble reports by customers. On the other hand, the resulting requirements which must be placed on the allowable transmission degradation of a given single link in the network may at first seem unreasonably stringent (out of context). One must remember that each link or part must introduce only a small portion of the total allowable degradation. With an objective of (for example) a 50-dB peak signal-to-rms-noise ratio for the end-to-end worst-case transmission path through the network, and allowing for realistic errors in the implementation of the network, it is not unusual to require a given link to have a nominal signal-to-noise ratio of 70 dB. Clearly if that link were the only link between the video source and the customer's receiver, the requirements could be relaxed significantly.

The loops between the video cameras and playback equipment (studios), and the broadcast stations, can be provided using cables containing shielded twisted copper pairs or using coaxial cables. When a studio is very far from the broadcast station (e.g., signals generated by television "networks" which are distributed to distant network subsidiaries) a microwave system can be used for the long-distance portion of the loop (with cable facilities interfacing the microwave system at each end). As mentioned, cable television distribution utilizes coaxial cable facilities. As discussed in Chapter 4, Section 4.3, the transmission of analog video signals at high signal-to-noise ratios over fiber systems is a difficult technical challenge. Therefore applications for analog fiber video transmission systems are presently limited to situations where the end-to-end transmission distances (signal-source-to-signal-sink) are short, or where the total allowed amount of signal degradation in transmission can be allocated heavily to the fiber system (at the expense of tightened requirements on the other links in the network). One feature motivation for using fibers is again the immunity to interference from electromagnetic fields of a small, lightweight, flexible, fiber cable.

For long-distance transmission of video signals over fiber cables, digital transmission is significantly more feasible (technically) than analog transmission. To prove-in fibers (over analog transmission on coaxial cables) the transmission distance must be long enough to save enough cost in transmission to pay for the analog-to-digital converters required at each end of the fiber system. Simple analog-to-digital converters, which can code 5-MHz analog video signals into 50–100-Mbaud digital signals, cost several thousand dollars at present. Technology advances may reduce these costs significantly. More complex analog-to-digital converters called "Frodecs" which require less than 10 Mbaud of digital transmission capacity presently cost 50–100 thousand dollars. A significant technical challenge, to make digital fiber optic distribution of video signals economically viable, is to reduce these Frodec costs by several orders of magnitude.

5.4. Interactive Video Distribution and Two-Way Video

In the classical entertainment video distribution, described above, each customer receives one or more video signals simultaneously, which are selected by the broadcaster or the cable television operator. The customer selects the desired program from the FDM signal using the tuner on his or her television receiver.

An alternative video distribution architecture has been experimented with recently, which provides customers access to large numbers of entertainment programs as well as educational material, and two-way video communication with other customers. An example of such a network is shown in Figure 5.4. Each customer is provided a private video "downlink" from a distribution office toward his or her premises. The customer is also provided a voice bandwidth uplink back toward the distribution office, and possibly a video uplink as well. Using the voice bandwidth uplink the customer can request the distribution office to transmit an available video program on his video downlink. This is implemented with video switching machines which must have the capability of interconnecting video trans-

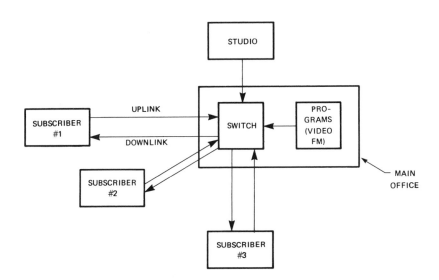

Figure 5.4. Interactive video network.

mission facilities in analog or digital form (whichever method is used). A customer with a two-way video loop may request to be interconnected with another such customer. Thus the possibility of two-way video communication is provided for. Using spare bandwidth or digital capacity on the downlinks, one or more entertainment audio signals can also be provided.

The considerations which determine whether coaxial cable systems or fiber systems are best suited for the transmission portions of this network are the same as described in Section 5.3 above. However, there is one additional feature-driven aspect. Since this architecture results in several hundred (or more) parallel video circuits, needed to provide the private channels, the relative size advantage of fiber cables becomes an important factor. A fiber cable containing several hundred fibers can be less than an inch in diameter. A cable with several hundred coaxial tubes would be several inches or several tens of inches in diameter.

5.5. Telemetry Links and Data Buses

Data buses were discussed in Section 4.2.2 above. They can take the form of point-to-point links or multiaccess networks. In either case the competitive technologies are coaxial cable; cables containing copper wire pairs; and (in the future) free space optical links.

The features which favor optical fibers include immunity to electromagnetic interference; a broad flat bandwidth; no fire hazard from sparks in the event of a cable cut; fibers are difficult to "tap" without being detected (fibers do not radiate significantly); terminals feeding redundant fiber paths are not shorted out when the cable containing one of the redundant paths is damaged; light weight and small size; and a low radar profile compared to wires.

The disadvantages of fibers compared to copper cable systems are the requirements for optical transmitters and receivers and for optical splices and connectors, which are relatively expensive when compared with their electrical counterparts in this type of application. Also, as mentioned in Chapter 4 in multiterminal systems, the access taps are relatively difficult to implement.

To date, many data bus applications have incorporated optical fiber links which can be directly substituted for wire pairs or coaxial cables in places where the terminals being interconnected were originally designed to work with the wire pair or coaxial systems. Thus the fiber links may have been at a disadvantage. The negative aspects of the differences between fibers and metallic cables have penalized the fiber systems, while the positive aspects of fibers relative to metallic cables may not have been utilized properly. The true test of the applicability of fibers to data buses and telemetry links will occur when networks consisting of both terminals and transmission systems are jointly optimized with fibers as one of the transmission alternatives.

5.6. Military Applications

Military applications for voice communication, video communication, and data transmission are similar to commercial applications, but with some added requirements. In military applications there is more emphasis on maintaining privacy, avoiding spurious radiation, and on small lightweight systems (for tactical applications). Here the features of optical fiber cables relative to wire pair and coaxial cables have increased importance.

One application, which is a caricature of the unique military requirements, is the fiber guided missile. What is needed here is a lightweight, high-strength, low-attenuation and broad-bandwidth (for video signals) transmission medium which can be unreeled from a moving missile. The alternative is wire, which appears to be inadequate for long-distance high-bandwidth transmission. Whether fibers will have adequate strength for this application (about 1 million psi is needed) and whether the unreeling process introduces any transmission degradations (e.g., increased loss) is under study.

Another unique and challenging military application is the towed array. Here a cable with pressure sensors attached to one end must be unreeled from a submerged submarine. Again the requirements are for light weight, low loss, large bandwidth, and high strength.

5.7. Passive and Active Sensing

Fiber cables can be used to return signals from various types of sensors and transducers. The obvious approach is to use the output of the transducer to modulate a light source whose output is then coupled to a fiber. However, there are some interesting alternative approaches. One approach would be to use a transducer whose output is already in the form of a light signal at an appropriate wavelength for transmission through the fiber. Another interesting approach is shown in Figure 5.5.

Here a passive (nonelectrical) transducer, in the middle of a fiber loop, presents a varying attenuation depending upon the value of a physical parameter being monitored. A light "probe" signal is generated at a terminal and the level of the returned (looped-back) light signal is used to monitor the parameter under observation. For example, the transmission loss of a

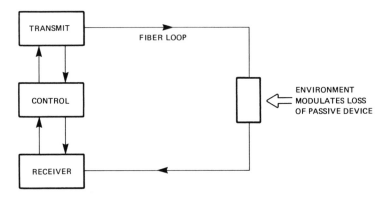

Figure 5.5. Passive looped sensing.

thin section of semiconductor material will vary with the temperature of the material (as the band gap of the material changes). Thus such a section of semiconductor can become a passive thermometer.

Figure 5.6 shows a "reflectometer" approach, where the passive sensor must have a reflection coefficient which varies with the parameter being monitored. For example, a flat unterminated fiber end would have a reflection coefficient of 4% in air. Such an unterminated end could be used to determine when a liquid in a container has reached a desired fill level. The contact of the liquid with the fiber end would modify the reflection coefficient, thus causing a change in the level of the returned (echo) signal.

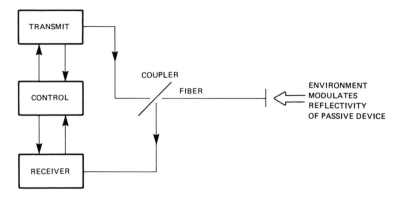

Figure 5.6. Passive reflectometer sensing.

References

Preface

1. Kao, K.C., and Hockham, G.A., Dielectric surface waveguides for optical frequencies, *Proc. IEE* **133**, 1151–1158 (July 1966).
2. Miller, S.E., Marcatili, E.A.J., and Li, T., Research toward optical fiber transmission systems, *Proc. IEEE* **61**, 1703–1751 (December 1973).
3. Sell, D., and Maione, T., Experimental fiber optic transmission system for interoffice trunks, *IEEE Trans. Commun.* **COM-25** (5), 517–522 (May 1977).
4. Atlanta Experiment, *Bell Syst. Tech. J.* **57**(6) Part 1 (July–August 1978).
5. Personick, S.D., Fiber optic communication, a technology coming of age, *IEEE Commun. Soc. Mag.* **16** (2), 12–20 (March 1978).

Chapter 1

1. Bell Labs Staff, *Transmission Systems for Communication,* Western Electric Company, Technical Publications, North Carolina (December 1971).
2. Miller, S.E., Marcatili, E.A.J., and Li, T., Research toward optical fiber transmission systems, *Proc. IEEE* **61**, 1703–1751 (December 1973).
3. Barnoski, M.K., in *Semiconductor Devices for Optical Communication,* Springer-Verlag, Berlin (1980), Chapter 6.
4. Miller, S.E., and Chynoweth, A.G., *Optical Fiber Telecommunications*, Academic Press, New York (1979), Chapters 3–15.
5. Kressel, H., *Semiconductor Devices for Optical Communications*, Springer-Velag, Berlin (1980), Chapters 2,7.
6. Sze, S.M., *Physics of Semiconductor Devices*, Wiley, New York (1967).
7. McIntyre, R.J., Multiplication noise in uniform avalanche photodiodes, *IEEE Trans. Electron Devices* **ED-13**, 164–168 (January 1966).
8. McIntyre, R.J., The distribution of gains in uniformly multiplying avalanche photodiodes, *IEEE Trans. Electron Devices* **ED-19**, 713–718 (June 1972).
9. Personick, S.D., New results on avalanche multiplication statistics with applications to optical detection, *Bell Syst. Tech. J.* **50** (1), 167–189 (January 1971).
10. Personick, S.D., Statistics of a general class of avalanche detectors with applications to optical communication, *Bell Syst. Tech. J.* **50** (10), 3075–3095 (December 1971).

Chapter 2

1. Bell Labs Staff, *Transmission systems for communications*, Western Electric Company, Technical Publications, North Carolina (December 1971).
2. Bell Labs Staff, *Engineering and Operations in the Bell System*, Bell Laboratories Inc., Murray Hill, New Jersey (1977).
3. Wozencraft, J.M., and Jacobs, I.M., *Principles of Communication Engineering*, Wiley, New York (1965).
4. Lucky, R., Salz, J., and Weldon, E., *Principles of Data Communication*, McGraw Hill, New York (1968).
5. Takasaki, Y. *et al.*, Optical pulse formats for fiber optic digital communications, *IEEE Trans. Commun.* **COM-24**, 404–413.
6. Hubbard, W., Efficient utilization of optical frequency carriers for low and moderate bit rate channels, *Bell Syst. Tech. J.* **52**, 731–765 (May–June 1973).
7. Sato, M., *et al.* Pulse interval and width modulation for video transmission, *IEEE Trans. Cable Television* **3**(4), 165–173 (October 1978).

Chapter 3

1. Personick, S.D., Receiver design for digital fiber optic communication systems, *Bell Syst. Tech. J.* **52**(6), 843–886 (July–August 1973).
2. Hubbard, W.M., Efficient utilization of optical frequency carriers for low to moderate bit rate channels, *Bell Syst. Tech. J.* **52**(5), 731–765 (May–June 1973).
3. Personick, S.D., Receiver design for optical fiber systems, *Proc. IEEE* **65**(12), 1670–1678 (December 1977).
4. Miller, S.E., and Chynoweth, A.G. (editors), *Optical Fiber Telecommunications*, Academic Press, New York (1979).
5. Barnoski, M.K. (editor), *Fundamentals of Optical Fiber Communications*, Academic Press, New York (1975).
6. Kressel, H. (editor), *Topics in Applied Physics*, Vol. 39, *Semiconductor Devices for Optical Communication*, Springer-Verlag, Berlin (1980).
7. Shumate, P.W., Chen, F.S., and Dorman, P.W., *Bell Syst. Tech. J.* **54**, 1823 (1978).
8. Van Trees, H.L., *Defection Estimation and Modulation*, Vol. I, Wiley, New York (1968).
9. Hullet, J.L., and Moui, T.V., A feedback receive amplifier for optical transmission systems, *IEEE Trans. Commun.* **COM-25**, 1180–1185 (October 1976).
10. McIntyre, R.J., Multiplication noise in uniform avalanche photodiodes, *IEEE Trans. Electron Devices* **ED-13**, 164–168 (January 1966).
11. Personick, S.D., New results on avalanche multiplication statistics with applications to optical detection, *Bell Syst. Tech. J.* **50**(1), 167–189 (January 1971).
12. Personick, S.D., Statistics of a general class of avalanche detectors with applications to optical communications, *Bell Syst. Tech. J.* **50**(10), 3075–3095 (December 1971).
13. McIntyre, R.J., The distribution of gains in uniformly multiplying avalanche photodiodes, *IEEE Trans. Electron Devices* **ED-19**, 713–718 (June 1972).
14. Personick, S.D., *et al.*, A detailed comparison of four approaches to the calculation of the sensitivity of optical fiber system receivers, *IEEE Trans. Commun.* **COM-25**, 541–548 (May 1977).
15. Cariolaro, G.L., Error probability in digital fiber optic communication systems, *IEEE Trans. Inf. Theory* **IT24** (March 1978).
16. Hauk, W., Bross, F., and Ottka, M., The calculation of error rates for optical fiber systems, *IEEE Trans. Commun.* **COM-26**(7), 1119–1126 (July 1978).
17. Papoulis, A., *Probability, Random Variables, and Stochastic Processes*, McGraw Hill, New York (1965).

18. Sell, D., and Maione, T., Experimental fiber optic transmission system for interoffice trunks, *IEEE Trans. Commun.* **COM-25**(5), 517–522 (May 1977).
19. Gardner, F., *Phaselock Techniques*, Wiley, New York (1979).
20. Duttweiler, D. L., The jitter performance of phase locked loops extracting timing from baseband data waveforms, *Bell Syst. Tech. J.* **55**, 37–58 (January 1976).
21. Runge, P. K., Phase frequency locked loops with signal injection for increased pull-in range and reduced output phase jitter, *IEEE Trans. Commun.* **COM-24**, 636–643 (June 1976).
22. Personick, S. D., Baseband linearity and equalization in fiber optical digital communication systems, *Bell Syst. Tech. J.* **52**(7), 1175–1194 (September 1973).
23. Personick, S. D., Time dispersion in dielectric waveguides, *Bell Syst. Tech. J.* **50**, 843–859 (1971).
24. Personick, S. D., and Barnoski, M. K., Measurements in fiber optics, *Proc. IEEE* **66**(4), 429–440 (April 1978).
25. Personick, S. D., Photon probe, an optical time domain reflectometer, *Bell Syst. Tech. J.* **36**(3) 355–366 (March 1977).
26. Barnoski, M. K., *et al.*, Optical time domain reflectometer, *Appl. Opt.* **16**, 2375 (1977).
27. Cohen, L. G., Shuttle pulse measurements of pulse spreading in an optical fiber, *Appl. Opt.* **14**, 1351 (1975).

Chapter 4

1. Barnoski, M. K. (editor), *Fundamentals of Optical Fiber Communications*, Academic Press, New York (1975), Chapter 3.
2. Kressel, H. (editor), *Semiconductor Devices for Optical Communication*, Springer-Verlag, Berlin (1980), Chapter 6.
3. Hubbard, W. M., Efficient utilization of optical frequency carriers for low and moderate bit rate channels, *Bell Syst. Tech. J.* **52**, 731–765 (May–June 1973).
4. Ito, T., *et al.*, Laser mode partition noise evaluation for optical fiber transmission, *IEEE Trans. Commun.* **COM-28**(7), (February 1980).
5. Miller, S. E., and Chynoweth, A. G. (editors), *Optical Fiber Telecommunications*, Academic Press, New York (1979), Chapter 4.
6. Miller, S. E., and Chynoweth, A. G. (editors), *Optical Fiber Telecommunications*, Academic Press, New York (1979), Chapters 6 and 13.
7. Atlanta Experiment, *Bell Syst. Tech. J.* **57**(6), Part 1 (July–August 1978).
8. Personick, S. D. (editor), *IEEE Trans. Commun., Special Issue on Fiber Optics*, **COM-26**(7), (July 1978).
9. Personick, S. D., Fiber optics communication, a technology coming of age, *IEEE Commun. Soc. Mag.* **16**(2),12–20 (March 1978).
10. Maione, T. L., and Sell, D. D., Experimental fiber optic transmission system for inter-office trunks, *IEEE Trans. Commun.* **COM-25**(5), 517–522 (May 1977).
11. Albanese, A., and Lensing, H., I.F. lightwave entrance links for satellite communications, Proc. ICC79, Paper I.7.1.–I.7.5 IEEE Publication No. 79 Ch1435-7, CSCB Boston (June 1979).
12. Personick, S. D., Receiver design for digital fiber optic communication systems, *Bell Syst. Tech. J.* **52**(6), 843–886 (July–August 1973).
13. Sato, M., *et al.*, Pulse interval + width modulation for video transmission, *IEEE Trans. Cable Television* **3**(4), 165–173 (Oct. 1978).

Chapter 5

1. Bell Labs Staff, *Transmission Systems for Communications*, Western Electric Company, Technical Publications, North Carolina (December 1971).

2. Bell Labs Staff, *Engineering and Operations* in the Bell System, Bell Laboratories Inc., Murray Hill, New Jersey (1977).
3. Personick, S.D., Digital transmission building blocks, *IEEE Commun. Soc. Mag.* **18**(1), 27–36 (January 1980).
4. Atlanta Experiment, *Bell Syst. Tech. J.* **57**(6) Part 1 (July–August 1978).

Index